A Guide to Native Orchids *of* South Western Australia

First published 2006.

R&R Publications Australia Pty Ltd

National Library of Australia
Cataloguing-in-Publication data:

Liddelow, W. R.

A guide to native orchids of south western Australia.

1st ed.
Includes index.
ISBN 0 9587532 4 5.

1. Orchids – Western Australia, Southwestern – Identification. 2. Endemic plants - Western Australia, Southwestern.
I. Title.

584.4099412

Photography: Bob Liddelow
Design: Scout Creative, scoutcreative.com.au
Print: Lamb Print, lambprint.net.au

Contents

Foreword

I have had an interest in orchids since childhood. Living near bush areas, I often used to wander into those areas seeking out spider orchids in particular. My interest has been regenerated in recent years to such an extent that I travel throughout the greater South-west at every opportunity, investigating different areas and searching for orchid species new to me.

In this time, I have accumulated some two thousand photographs of native orchids. This has enabled me to present 142 different orchids, adequately illustrated, in this book.

Because I have been successful most often when directed to a particular area, and have mostly been frustrated at blind 'bush-bashing', I decided to write this book. I trust that many will benefit from it, especially from the way in which I have cross-referenced the selected orchids and specific sites.

The reader should understand that I have described, illustrated, and given a location for, only those orchids which I have actually found and photographed myself. There are many omissions from those which exist because I have not photographed them sufficiently well, or they have not yet been located by me.

In fact, in this book, the 142 orchids I have included represent less than half of those now recorded in W.A.

To make best use of this book, I would recommend that the reader first read through the Author's Notes and the Glossary. Then skim through the text and mud maps to familiarise yourself with the organisation I have adopted. Then you will be ready to start, or continue, your adventure with some of Australia's most unique plants.

If you live in the Metropolitan area and are just starting, I would recommend that you begin in Kings Park or Bold Park looking for the Carousel Spider Orchid and other spider orchids. Also, several donkey orchids should be readily discovered.

Author's Notes

The family ORCHIDACEAE is huge, and consists world-wide of more than 700 genera which, in turn, comprise in excess of 25,000 species. The bulk of these are epiphytes living in tropical and sub-tropical regions.

The south-west of Western Australia has only terrestrial native orchids, but with around 400 such species it has a high proportion of the terrestrial orchids found in Australia.

Orchids are distinguished from other flowering plants by their unique column, which is a structure in which the stamens and pistils are fused together instead of being separate. They are generally characterised by having three petals and three sepals. Usually one of the petals is modified into a labellum, or lip, which is basically to attract insects for pollination. Thus many of the orchids use colour and scents to attract a variety of wasps, flies, bees, gnats, and ants.

In this book I have described only those orchids which I have actually found and photographed myself. For this, I have selected, for each species, the best photograph in my collection to show the main features of that orchid. The description generally emphasises the probable distribution of an orchid, and details on its flower. I have only occasionally referred to leaves, because most fall into a pattern: i.e. *Caladenias* mostly have a simple, long, flat, narrow, hairy leaf. Leek and mignonette orchids have a long tubular leaf. Hammer and duck orchids have a small, heart-shaped leaf. *Pterostylis* species usually have a rosette of small leaves.

Similarly, the sites noted for each orchid are ones where I have located the relevant species on at least one occasion, generally many times. There is no guarantee, of course, that these orchids will be found at the same sites in following seasons. The effects of fire, drought, flood, animal activity, orchid enthusiasts' activity, non-fertilization, and so forth can cause orchids to relocate next season, or disappear altogether. However, at least you now have some guidance to focus on a localised area, which is far better than blind 'bush bashing', without any real leads.

I have cross-referenced all the orchids and all the sites for your convenience.

My sites are represented by 'mud maps', which are as precise as I can make them. They are localised in arbitrary regions, decided by me.

These are:

SW – South West
SC – South Central
SE – South East
E – East
N – North

These are shown on the inside back cover of the book.

From this, I should point out several purposes for this book:

- To provide likely locations for specific orchids that the reader is seeking.
- To help identify orchids found, by comparing them with the photographs and descriptive text.
- To provide the location of sites when planning a trip.
- To provide a challenge to find more orchids.

For Rare Flora, I have kept specific sites confidential. As most of the orchids in this category are classified as Threatened, it is prudent to not encourage people other than researchers, conservationists and the like to visit such sites. However, if the reader locates one of these species, the orchid researchers at the Department of Environment and Conservation will be interested to hear of it. Similarly, I have not included any sites which are on private land.

Please note that 'My Best Time' for finding each orchid relates to my experiences, and usually covers a fortnightly or monthly period.

However, many orchids are listed in other publications as flowering over much greater periods, often starting well before 'My best time' and/or finishing well after. Again, I am providing the period of my most common discoveries and hopefully, then, your most likely period of finding particular orchids. However, do not disregard the possibilities of finding an orchid before, or after, 'My Best Time'. Weather can significantly affect the flowering season, particularly the rainfall pattern, and the geographic range of each orchid.

The order of the book has *Caladenias* first, because they have the greatest number of species, and they were my childhood favourites, especially the spider orchids. These have been arranged roughly on the basis of mostly white flowers through to coloured species, and then the more unusual shapes. The other genera follow.

I have supplied both scientific and common names for each species, and indexes for both. In the text I have used common names, assuming that the reader is more of an enthusiast, than a professional botanist. Note also, that scientific names are currently changing, as classification becomes more accurate with the use of new techniques, such as DNA sequencing, and re-examination of earlier classifications. Many of the orchids listed as aff. *species* have proper names in manuscript form but have not been officially named as yet.

The reader should note that I have not specified sites within the metropolitan area because they are too numerous, and I leave it as a challenge to go orchid hunting around Perth. Notably Kings Park and Bold Park have a good range of orchids, and I would recommend a recent publication:

Perth Plants – a field guide to the bushland and coastal flora of Kings Park and Bold Park, Perth, Western Australia
Burnett, Russell L and Tay, Eng Pin
Botanic Gardens and Parks Authority (W.A.), 2005

There are many other Nature Reserves and Conservation areas to be investigated, such as Wireless Hill, Star Swamp, Warwick Open Space, Bibra Lake, John Forrest National Park, Whiteman Park, Hollywood Reserve, etc. Even remnant bush, road verges, and golf courses can be worthwhile exploring.

Also note that I have not yet discovered the Elbow Orchid, the Pygmy Orchid, the Underground Orchid, the Potato Orchid, the Babe-in-a-cradle, any Helmet orchids, nor any Beard orchids. All other Western Australian orchid genera have been represented, albeit incompletely.

Please take care when using this book in the field:

- To be aware of natural dangers, such as snakes, spiders, poisonous plants, spiky undergrowth, deep holes, etc.
- To never pick orchids; simply admire them and/or photograph them.
- To be careful not to trample other orchids while moving around to photograph or spot others.
- To explore the area around my specific sites, as orchids can shift location, and other unlisted species may be found nearby.
- To check out burned areas; many orchids flower best after summer fires – but don't be surprised to find nothing.
- To leave nothing behind, especially this book, glasses, hats, drink bottles, etc.

Acknowledgements

First and foremost, I would like to acknowledge the enormous help and encouragement of my wife, Rosemary. She has far better 'eyes' for spotting orchids than I have, and has often pushed on when I'm ready to give up. She has been invaluable on the many trips we have made as 'orchid enthusiasts'.

I would also sincerely thank Andrew Brown for his generous assistance in identifying my 'finds', and providing many leads to areas and species new to me. He has been an inspiration to me for several years in many ways. Without Andrew, this book would never have been written. I also thank him wholeheartedly for finding time in his busy work schedule to check the manuscript for errors of mine.

Others who have offered guidance in finding, and identifying orchids, and to whom I am extremely indebted include Greg Bussell, Steve Hopper, Ken Mulavin, Judy Henderson, and Margaret Winchcombe.

Relating more to the development of this book, I would also like to acknowledge the friendly and professional photographic assistance of Domenic Papaelia and Diana Kelly of The Camera and Photo Specialists, Wembley.

For the layout and design, I marvel at the meticulous and creative way in which Steve Horton of Scout Creative has put it all together.

As my consultant at Lamb Print Pty Ltd, the enthusiastic Lianne Spears has been encouraging and ever-willing to help me get everything right. Thanks to Brett Todhunter for his artistry in enhancing the photos in such a way as to produce the colours and contrasts which show the orchids in their peak condition.

As usual, my business partner, Gary Dinsdale has taken on the task of typing and formatting my scribbled manuscript notes. He has patiently changed what was already done every time I had a new thought. He has been most gracious in letting me have my own way on this book.

My thanks also to my long-time friend Ken Quigley who has painstakingly helped proof read the manuscript to ensure that my oversights are minimised.

For an authoritative and comprehensive book on all W.A. orchids, get a copy of

Orchids of South-west Australia
Hoffman, Noel and Brown, Andrew
University of Western Australia Press, Nedlands, W.A.
Revised second edition, with supplement, 1998

For a more technical publication, on mostly *Caladenia*, I would recommend

Nuytsia, Vol 14 number 1/2 2001
Western Australian Herbarium, CALM, W.A.

For a Web-based reference, go to
http://florabase.calm.wa.gov.au

Finally, I would invite comment and information from readers.

Please ring me on (08) 9385 8069, or FAX (08) 9385 8064, or

write to me at 6 Yanagin Crescent, City Beach, W.A. 6015, or

email roandboblid@bigpond.com

I trust you will enjoy hunting orchids as much as I have enjoyed working on this book.

WR (Bob) Liddelow
City Beach, W.A.
August 2006

The Orchids

Caladenia longicauda subsp. *longicauda*

White Spider Orchid

Relative Caladenia Flower Size: large

The White Spider Orchid is similar to many of the other white flowered spider orchids, but has a much larger labellum than many, with up to 8 rows of red calli, and a generally taller fringe. The slender petals and sepals are stiffly held and may then hang downwards.

It is widespread in the *Banksia* and Jarrah woodlands from Lancelin to Nannup and Mt. Barker.

My site: Map **4**, SW 5

My best time: October

Caladenia longicauda subsp. *calcigena*

Coastal White Spider Orchid

Relative Caladenia Flower Size: large

Found on the Swan coastal plain from 30 km south of Dongara to Bunbury, this large white spider orchid prefers limestone-based soils in low scrub associated with *Banksia* and Tuart woodlands.

Like most of its genus, it has long, pendulous petals and sepals, but has quite small red calli that are not always regularly arranged.

My site: Map **43**, N 1

My best time: September

Caladenia longicauda subsp. *clivicola*

Hill's White Spider Orchid

Relative Caladenia Flower Size: large

This orchid is very similar to the Coastal White Spider Orchid, but grows on the Darling Scarp near Harvey and along the Leeuwin-Naturaliste Ridge. It prefers Marri and Jarrah forests.

Structurally it differs from the Coastal White Spider Orchid in having a smaller labellum with regular rows of calli and short fringes. The flowers are usually less brilliantly white.

My site: Map 5, SW 7

My best time: early October

Caladenia longicauda subsp. *eminens*

Stark White Spider Orchid

Relative Caladenia Flower Size: large

This orchid is aptly named, and is distinguishable from similar species by its brilliant white, broader petals and sepals, and broad labellum.

It is often found in spectacular large colonies in Wandoo and Yate woodlands. It grows in a triangular area of wheatbelt country, from Moora down to Cranbrook and across to Esperance.

My sites:

Map **24**, SC 7
Map **25**, SC 9,10
Map **26**, SC 11
Map **28**, SE 1,2
Map **29**, SE 3
Map **30**, SE 5
Map **32**, SE 9
Map **39**, E 2

My best time:

October

Caladenia longicauda subsp. *crassa*

Esperance White Spider Orchid

Relative Caladenia Flower Size: very large

This white spider orchid is restricted to areas north and east of Esperance, favouring swampy and rocky terrain.

The base of each petal and sepal is relatively wide, the labellum is moderately large, with relatively short fringes. The red calli are in rows of 4 to 8.

My site: Map **34**, SE 14

My best time: late September

Caladenia longicauda subsp. *rigidula*

Rigid White Spider Orchid

Relative Caladenia Flower Size: large

This orchid is uncommon and restricted to granite outcrops and wet areas north-east of Esperance.

The petals and sepals tend to be stiffly held and then are partly downcurved. The fringes are relatively short.

My site: Map **34**, SE 14

My best time: September

Caladenia longicauda subsp. *borealis*

Daddy-long-legs Spider Orchid

Relative Caladenia Flower Size: large

This white spider orchid has flowers with stiffly held petals and sepals. The labellum is relatively narrow with long fringes. It often grows in clumps, especially in sandy-clay soils between Kalbarri and Cataby. Those pictured above were next to the beach at Lake Indoon.

Their other habitat is mainly Wandoo woodland, Sheoak areas and *Acacia* thickets.

My sites: Map **47**, N 5 Map **48**, N 9, 10

My best time: August

Caladenia longicauda subsp. *merrittii*

Merritt's White Spider Orchid

Relative Caladenia Flower Size: very large

This is an impressive white spider orchid found in Jarrah forest between Augusta and Nannup.

The sepals and petals hang almost vertically and the dorsal sepal usually curves backwards and downwards.

I have often found it as solitary plants in the Alexandra Bridge area, but it can be easily confused with the Splendid White Spider Orchid which is more widespread. The latter has a very much larger labellum, higher fringes, and larger calli.

My sites: Map **11**, SW 21, 38, 39; Map **14**, SW 32; Map **16**, SW 40; Map **19**, SW 43, 44

My best time: early October

Caladenia splendens

Splendid Spider Orchid

Relative Caladenia Flower Size: very large

This orchid has the largest flowers of the white spider orchids and usually stands much taller. The creamy-white petals and sepals hang almost vertically. The large labellum has a high maroon fringe with white tips, and there are usually 4 rows of tall, pale maroon calli.

The Splendid Spider Orchid dwells in wet areas of Jarrah forest in a triangular region from the Avon Valley National Park south to Nannup and Mt Barker.

My sites: Map **11**, SW 21, 39 Map **19**, SW 43 Map **22**, SC 3, 5

My best time: late October

Caladenia lodgeana

Lodge's Spider Orchid

Relative Caladenia Flower Size: large

This spider orchid is easily confused with the Christmas Spider Orchid in the Augusta region where it grows, often in the same area as the Christmas Spider Orchid. It is distinguishable from the latter by its more stiffly pointing petals and slightly clubbed sepals. The flowers are usually smaller.

It is confined to winter-wet, clay-based soils near Augusta.

My sites: Confidential

My best time: mid November

Caladenia serotina

Christmas Spider Orchid

Relative Caladenia Flower Size: large

In the Augusta area, I have found some difficulty in distinguishing this orchid from Lodge's Spider Orchid. However, the Christmas Spider Orchid tends to flower later than Lodge's Spider Orchid, and its petals tend to hang down like the lateral sepals.

The labellum is larger with the tall fringe almost vertical.

The flowers range from plain white to having yellow and red suffusions.

It is found scattered widely in a triangular area between Perth, Augusta and Albany.

My site: Map **13**, SW 26 a

My best time: late November

Caladenia sp. Rothsay

Rothsay Spider Orchid

Relative Caladenia Flower Size: large

This spider orchid occurs in an area extending 50-100 km either side of the Vermin Proof Fence from above Perenjori to Mullewa. This orchid prefers rocky ground in shrublands.

It can be confused with the Glistening Spider Orchid (*Caladenia incensa*) which extends further north and east, and has glistening white flowers and a broader leaf. The Rothsay orchid has pale yellow, stiffly held sepals and petals that are held horizontally then downcurve vertically. The dorsal sepal is held upright and slightly incurved. I have found it growing in clumps, rather than as single plants.

My site: Map **51**, N 17

My best time: late August

Caladenia christineae

Christine's Spider Orchid

Relative Caladenia Flower Size: large

This attractive spider orchid is usually found as single plants near winter-wet areas and lakes in small pockets between Mt Barker and Manjimup. It is associated with Jarrah, Marri, and Paperbark areas.

It is declared as Rare Flora.

My sites: Confidential

My best time: early October

Caladenia remota subsp. *remota*

Outback Spider Orchid

Relative Caladenia Flower Size: large

This white spider orchid is found in an area in the north-east of the greater south-west in a region embracing Wubin, Morawa, Canna, and Yalgoo. It prefers moist-wet areas.

It is characterised by a relatively small labellum and stiffly splayed, tapering petals and sepals.

My site: Map **51**, N 16

My best time: August

Caladenia vulgata

Common Spider Orchid

Relative Caladenia Flower Size: large

This spider orchid is widespread in a large triangle over the greater south-west. This extends from Kalbarri to Esperance, then west to Denmark. It is not found in the Capes region, nor the coastal areas to Denmark.

It favours sandy soils, often grows in clumps, and shows considerable variation in the white petals and sepals. The latter are distinctively pendulous.

My sites: Map **3**, SW 3, 4
Map **26**, SC 11
Map **46**, N 4

My best time: August and September

Caladenia nobilis

Noble Spider Orchid

Relative Caladenia Flower Size: very large

This orchid is similar to the Common Spider Orchid (*C. vulgata*), but has larger flowers and a wider labellum. It shows similar variations in colour.

Its habitats also vary throughout its distribution that covers a broad inland band between Kalbarri and Bunbury, incorporating Sheoak, Peppermint and Tuart woodlands.

My sites: Map **3**, SW 4; Map **44**, N 2; Map **49**, N 12

My best time: September

Caladenia dimidia

The Chameleon Orchid

Relative Caladenia Flower Size: large

This orchid is found mainly in Mallee scrub, especially near winter-wet creeks and granite outcrops in the eastern wheatbelt from Payne's Find to Esperance.

The floral parts are usually cream-white with maroon variations. The petals tend to be held horizontally.

My sites: Map **36**, SE 17, 18
Map **37**, SE 19
Map **52**, N 18
Map **53**, N 20
Map **54**, N 21

My best time: early September

Caladenia polychroma

Joseph's Spider Orchid

Relative Caladenia Flower Size: large

This is another spider orchid displaying a wide range of colours from white through yellow, orange and red. (Hence its name)

It is found in a triangular area between Mt Barker, Kojonup, and Hopetoun, in Yate and Wandoo woodlands.

My sites: Map **25**, SC 8, 9, 10 Map **26**, SC 11

My best time: early October

Caladenia microchila

Western Wispy Spider Orchid

Relative Caladenia Flower Size: large

This spider orchid is found most abundantly in mallee woodlands, but also in coastal limestone-based soils. It has long, hanging, cream-white petals and sepals that often become tangled with those of adjacent flowers. The central floral parts are quite small.

It grows in a large region from Hyden and towards the Great Australian Bight, and towards the coast around Hopetoun.

My site: Map **33**, SE 13

My best time: September

Caladenia busselliana

Bussell's Spider Orchid

Relative Caladenia Flower Size: medium large

An attractive, but elusive orchid, with pale yellow petals and sepals, the latter having long narrow clubbing. The labellum is similarly coloured with dark maroon, fairly long fringes. The dorsal sepal is usually erect and slightly curved forwards.

It is declared as Rare Flora, and is found in the general Yallingup area in a few winter-wet areas of Marri woodland.

My sites: Confidential

My best time: early October

Caladenia pholcoidea subsp. *pholcoidea*

Albany Spider Orchid

Relative Caladenia Flower Size: very large

This stunning orchid is readily identified by its pale cream-yellow flower, relatively small, shortly fringed labellum, its considerable height, and most of all, its late flowering period through November and December.

It is found around winter-wet areas such as swamps and creeks in Jarrah forest. Its distribution is from Albany to Augusta.

My sites: Map **13**, SW 24, 25, 27

My best time: late December

Caladenia citrina

Margaret River Spider Orchid

Relative Caladenia Flower Size: large

This attractive spider orchid has pale lemon coloured floral parts with distinctive clubbed fringes. The four rows of calli are usually maroon towards the base of the labellum.

It grows in Jarrah-Marri woodlands of the Leeuwin-Naturaliste Ridge from Dunsborough to south of Witchcliffe.

My site: Map **11**, SW 19

My best time: late September

Caladenia caesarea subsp. *caesarea*

Mustard Orchid

Relative Caladenia Flower Size: very small

The Mustard Orchid is mainly found in Wandoo woodlands in a belt between Boyup Brook and Cranbrook.

It has relatively small flowers with an attractive yellow coloration and red-brown markings. The petals and sepals are stiffly held and the relatively large labellum protrudes outwards and downwards at the apex.

My sites: Map **21**, SC 2 Map **25**, SC 8

My best time: late October

Caladenia caesarea subsp. *maritima*

Cape Spider Orchid

Relative Caladenia Flower Size: small

The flowers of this orchid are a very distinctive yellow with red-brown stripes on the labellum.

It usually grows in clumps, restricted to granite areas near Dunsborough and the Ludlow Tuart forest. It is declared as Rare Flora.

My sites: Confidential

My best time: September

Caladenia chapmanii

Chapman's Spider Orchid

Relative Caladenia Flower Size: large

A sandy soil dweller, Chapman's Spider Orchid has flowers that exhibit a range of colours, predominantly maroon with red, yellow and white.

It is mainly found between Bunbury and Margaret River among Peppermint woodlands, with a separate region further inland in Wandoo areas. It is often found growing in clumps.

My sites: Map **5**, SW 6, 7
Map **6**, SW 10
Map **10**, SW 17

My best time: early October

Caladenia footeana

Crimson Spider Orchid

Relative Caladenia Flower Size: medium large

This spider orchid is similar to others of this colouring, but has a distinctive structure, in that the petals and sepals are stiffly held and then curve downwards. The dark pink-red to maroon sepals and petals often become creamy-white towards their bases. The labellum has reddish basal lines and two rows of pink-red calli. The plants are relatively short.

The Crimson Spider Orchid favours wetter areas in a belt from Northampton to Cranbrook, particularly in Wandoo woodlands and dense thickets.

My Site: Map **26**, SC 11

My best time: late September

Caladenia pulchra

Slender Spider Orchid

Relative Caladenia Flower Size: large

The flowers of this orchid, although often entirely red, do exhibit yellow and white colour variations.

It is an uncommon orchid found most often in Sheoak woodlands near creeks in the southern wheatbelt as far south as Jerramungup.

My site: Map **32**, SE 9

My best time: early October

Caladenia harringtonae

Pink Spider Orchid

Relative Caladenia Flower Size: small

This attractive pink spider orchid may be all pink, but often has white margins to the petals and sepals. These are held stiffly horizontal before falling off into a hanging posture. The fringes are relatively short.

It is found in winter-wet flat areas adjacent to lakes and creeks in the Manjimup-Albany region. It is currently declared as Rare Flora.

My sites: Confidential

My best time: late October

Caladenia radialis

Drooping Spider Orchid

Relative Caladenia Flower Size: small

This is a very widespread orchid, ranging through the wheatbelt from Northampton to the Fitzgerald National Park. It has a wide variety of habitats.

The dorsal sepal is usually drooping (backwards), much like the petals and sepals. The red calli are fairly densely displayed on the labellum.

My sites: Map **48**, N 7, 8

My best time: early September

Caladenia drummondii

Winter Spider Orchid

Relative Caladenia Flower Size: very small

This orchid grows to only 8 cm in height and is usually found in thick leaf litter in Mallee woodlands. These factors, together with its deep maroon-brown colouring and very small flower size, make it very difficult to find.

The petals are often curled at the ends, which leaves its identity in no doubt.

It occurs throughout the wheatbelt from near Kalbarri down to Lake Grace.

My site: Map **40**, E 3

My best time: late May

Caladenia heberleana

Heberle's Spider Orchid

Relative Caladenia Flower Size: large

This spider orchid has a limited distribution in sandy soils in a narrow band from Albany and the Stirling Ranges east to Cape Arid.

It has pale yellow petals and sepals with variable maroon suffusions, a relatively large labellum having a maroon tip and long, white tipped fringes. The sepals and petals are slightly clubbed. It occurs in Jarrah, Marri and *Banksia* woodlands.

My sites: Map **34**, SE 15 Map **35**, SE 16

My best time: late September

Caladenia gardneri

Cherry Spider Orchid

Relative Caladenia Flower Size: medium large

This attractive orchid has a range of pink coloration, as illustrated, with the labellum invariably deep pink on the apex. The fringes are also deep pink and slanted forwards. The petals and sepals are stiffly held, with the sepals having slender clubbing.

It favours Peppermint woodlands in near-coastal consolidated dunes along the Leeuwin-Naturaliste Ridge and easterly along the southern coast to William Bay.

My site: Map **6**, SW 9

My best time: mid October

Caladenia nivalis

Exotic Spider Orchid

Relative Caladenia Flower Size: small

This orchid is striking when it is found in dark green, low, coastal vegetation. It is usually brilliantly white with the labellum apex a contrasting dark red. The widely splayed sepals and petals may be slightly downcurved. The sepals are narrowly clubbed, often a light brown in colour. The fringes are coloured and relatively short.

Its distribution is extremely limited to coastal granite outcrops, between Cape Naturaliste and Moses Rock.

My site: Map **6**, SW 8

My best time: late September

Caladenia applanata subsp. *applanata*

Broad-lipped Spider Orchid

Relative Caladenia Flower Size: large

The common name of this orchid describes its main distinguishing feature – a relatively large, broad labellum, which is dark maroon-red with similarly coloured, short fringes. The petals and sepals are also maroon-red with cream-coloured suffusions.

It tends to grow in coastal Peppermint woodlands from Cape Naturaliste to Albany.

My site: Map **7**, SW 11

My best time: late September

Caladenia viridescens

Dunsborough Spider Orchid

Relative Caladenia Flower Size: small

This is another of the spider orchids declared as Rare Flora, and its existence is threatened. Its main difference to similar orchids, such as Bussell's Spider Orchid, is the relatively large, deeply coloured labellum. The petals and sepals have more of a greenish tinge, and the labellum fringes are longer.

It favours well-drained, sandy soils in small areas of Cape Naturaliste.

My sites: Confidential

My best time: late September

Caladenia paludosa

Swamp Spider Orchid

Relative Caladenia Flower Size: small

This orchid is structurally very similar to the Dunsborough Spider Orchid but is not as greenish in colour and has shorter fringes.

Its habitat is more that of winter-wet areas between Gingin and Gracetown, and is relatively common. It generally flowers later.

My sites: Map **5**, SW 6, 7
Map **6**, SW 10
Map **7**, SW 12

My best time: early October

Caladenia arenicola

Carousel Spider Orchid

Relative Caladenia Flower Size: large

This is one of the most common and attractive spider orchids on the coastal plain, ranging from Lancelin to Yarloop. It prefers sandy soil (Bassendean sands) and grows mainly in Jarrah, *Banksia*, and Sheoak woodlands, particularly in Kings Park and Bold Park and other metropolitan nature reserves.

It displays a range of colours (as illustrated), usually with a red apex to the relatively large labellum. The fringes are relatively high and the calli prominently displayed. The sepals and petals are usually slightly clubbed and held stiffly.

The lutea (yellow) form illustrated below is a good example of particular plants appearing and disappearing from a site. I recorded three separate clumps of this in early September, every year from 1997 to 2000, but haven't seen it since.

My sites: Metropolitan areas such as Bold Park, Kings Park, Warwick Open Space, etc.

My best time: September

Caladenia georgei

Tuart Spider Orchid

Relative Caladenia Flower Size: large

This orchid is similar to the Carousel Spider Orchid, but has longer fringes, has no clubbing on the petals, and inhabits Spearwood sands between Yanchep and Busselton. It is usually found in Banksia and Peppermint woodlands as well as the Tuart forests for which it is named.

My site: Map **3**, SW 4

My best time: late September

Caladenia thinicola

Scott River Spider Orchid

Relative Caladenia Flower Size: large

This orchid is very similar to the Grand Spider Orchid that is declared as Rare Flora, and confined to a few patches on the coastal plain between Perth and Yallingup.

However, the flowers of the Scott River Spider Orchid are a lot smaller and a little shorter. Both have very long fringes and pale greenish petals and sepals, with the latter slenderly clubbed. There are usually maroon suffusions.

The Scott River Spider Orchid is found mainly in the Scott River National Park, but also in consolidated sand dunes further east along the coast to south of Pemberton. It is also found along the coast up to Yallingup. It is associated with Peppermint, Jarrah and Marri woodlands.

My sites: Map **6**, SW 9, 10; Map **7**, SW 12; Map **16**, SW 40

My best time: early October

Caladenia brownii

Karri Spider Orchid

Relative Caladenia Flower Size: large

This attractive spider orchid is found in the lower south-west, from Busselton towards Albany. As well as being found in Karri forest, it inhabits *Banksia*, Sheoak woodlands and coastal scrub.

The labellum is deeply coloured and almost smooth, with high fringes. This is one of the few spider orchids that have slender clubbing on the petals as well as the sepals. These flower parts are held stiffly.

My sites: Map **5**, SW 6
Map **12**, SW 22
Map **13**, SW 26 b
Map **19**, SW 43
Map **23**, SC 6 b

My best time: mid October to mid November

Caladenia ferruginea

Rusty Spider Orchid

Relative Caladenia Flower Size: medium large

This is one of my favourite spider orchids, having an almost smooth brownish-red apex to the otherwise creamy white labellum, upswept petals, an erect, incurved dorsal sepal, striped brownish-red sepals and petals, and distinctive light brown clubbing on the sepals.

It is commonly found in Jarrah, Marri, and coastal Peppermint woodlands in the Busselton-Margaret River region, but also extends northwards towards Perth and eastwards to the Porongurup Range.

My sites: Map **5**, SW 6, 7; Map **6**, SW 9, 10; Map **10**, SW 17; Map **11**, SW 18; Map **13**, SW 28

My best time: late September, early October

Caladenia pectinata

King Spider Orchid

Relative Caladenia Flower Size: large

This is another attractively coloured spider orchid, with yellowish petals and sepals, often with variable suffusions of pink. The large labellum has a dark maroon apex, with dark maroon to golden fringes. The sepals are predominantly clubbed, light brown in colour.

It is mostly found between the Fitzgerald National Park and Rocky Gully, with outlying populations as far north as Gingin and as far west as Boyup Brook. It prefers sandy, clay loams in wet areas.

My site: Map **23**, SC 6 a

My best time: October

Caladenia procera

Carbunup King Spider Orchid

Relative Caladenia Flower Size: large

This is another attractive spider orchid, especially because its tall scapes often reach nearly a metre in height. The sepals and petals are greenish-yellow, occasionally with maroon to pink suffusions. The dark maroon apex on the labellum and the tall maroon to lemon fringes make a good contrast. The sepals are prominently clubbed, being yellow to brown in colour. The petals are held stiffly and ascending.

It is confined to an area to the south-west of Busselton, favouring sand and loam in mixed Jarrah, Marri, and *Banksia* woodlands. This spider orchid is now declared as Rare Flora.

My site: Confidential

My best time: early October

Caladenia decora

Esperance King Spider Orchid

Relative Caladenia Flower Size: large

This is a very attractive orchid found to the east (Cape Arid) and west (Fitzgerald River National Park) of Esperance. It favours winter-wet swampy areas.

It is richly coloured with long fringes and a broad, deep maroon labellum. The sepals and (usually) the petals have yellow-brown clubs. The petals are sometimes downcurved and sometimes upcurved.

My sites: Map **34**, SE 15 Map **35**, SE 16

My best time: late September

Caladenia infundibularis

Funnel-web Spider Orchid

Relative Caladenia Flower Size: large

This is often a common orchid in the Augusta region, mainly as single plants. The flowers are a greenish yellow with a very dark maroon labellum apex. The latter is smooth and often folded to form a channel. This large labellum is nearly flat at the centre so that the fringes are close to horizontal. The petals are usually stiffly held above the horizontal. The sepals bear yellow-brown clubs.

It has a range of habitats, including granite outcrops, Jarrah and Karri forest, and coastal heath. It often grows up through ground level Balga.

My sites: Map **7**, SW 11; Map **12**, SW 22; Map **13**, SW 23, 26 b; Map **14**, SW 32

My best time: October

Caladenia uliginosa subsp. *uliginosa*

Darting Spider Orchid

Relative Caladenia Flower Size: large

This spider orchid is characterised by the petals and sepals downcurving from the base, whereas most other spider orchids have them joined horizontally and then curve downwards at some distance from the base.

The red-coloured labellum apex, pale green-cream petals and sepals, and high central parts also make it distinctive.

It is mainly found in Wandoo country south of York, then down as far as Rocky Gully.

My site: Map **22**, SC 5

My best time: late October

Caladenia excelsa

Giant Spider Orchid

Relative Caladenia Flower Size: very large

In full bloom, the Giant Spider Orchid is an imposing orchid. It is the largest of any of our spider orchids. It also stands tall with a scape of nearly a metre not uncommon.

The lengthy sepals and petals trailing in the wind make for an outstanding find in light scrub.

It is found in sandy soils in *Banksia* woodland, integrated with Marri and Jarrah. Its distribution from Yallingup south along the Leeuwin-Naturaliste Ridge is sparse and it is currently declared Rare Flora.

My sites: Confidential

My best time: early October

Caladenia radiata

Ray Spider Orchid

Relative Caladenia Flower Size: small

This orchid is readily recognised by its long, radiating fringes and projecting lateral sepals. The deep maroon suffusions on the narrow sepals and petals, and the deep maroon labellum with very pale fringes also make it readily identifiable.

It favours wet areas in scrub between Yarloop and Albany.

My site: Map **22**, SC 5

My best time: early November

Caladenia rhomboidiformis

Diamond Spider Orchid

Relative Caladenia Flower Size: small

As its name implies, this orchid has a distinctively shaped labellum. This is pale cream with a deep maroon apex. The fringes are very short, with relatively thick marginal calli. The petals are thin and tapering and usually held horizontally to downcurving. The lateral sepals are broad for the basal half, then very narrow, ending in yellow-pale brown clubbing. The petals and sepals are normally pale yellow with variable suffusions of maroon.

The main distribution is in the Busselton, Margaret River, Augusta strip. Its habitats vary, ranging from *Banksia* and Sheoak woodlands to Jarrah forests.

My site: Map **13**, SW 26 b

My best time: late September, early October

Caladenia ensata

Stumpy Spider Orchid

Relative Caladenia Flower Size: small

The Stumpy Spider Orchid is fairly readily recognised when first seen due to its small size, and disproportionately shortish sepals when compared with the Clubbed Spider Orchid. Both have thick yellow clubs.

The fringes are extremely short and the labellum is dark maroon with crowded dark maroon calli. The petals and sepals are pale yellow with variable maroon suffusions.

This orchid favours deep sandy soils in *Banksia* and Jarrah woodlands between Albany and Karridale. It is not commonly found.

My site: Map **16**, SW 40

My best time: October

Caladenia longiclavata

Clubbed Spider Orchid

Relative Caladenia Flower Size: small

This orchid is similar to the preceding Stumpy Spider Orchid but is distinguishable from it in having slightly larger flowers with much longer labellum fringes. The lateral sepals are proportionately larger, as are the clubs they bear.

The distribution of the Clubbed Spider Orchid is also greater – between New Norcia and Albany, including all the lower south-west.

It favours sandy soils in *Banksia* and Sheoak woodlands, as well as Jarrah and Karri forests.

My sites: Map **11**, SW 38 Map **14**, SW 32

My best time: early October

Caladenia magniclavata

Big Clubbed Spider Orchid

Relative Caladenia Flower Size: small

This orchid is very similar to the Clubbed Spider Orchid but, as its name implies, it has much longer clubs. The labellum is similarly dark maroon with crowded calli, but the latter are shorter in the Big Clubbed Spider Orchid.

It is usually found in heavy, sandy soils in Jarrah and Karri forests between Albany and Karridale.

My site: Map **11**, SW 38

My best time: early October

Caladenia arrecta

Reaching Spider Orchid

Relative Caladenia Flower Size: small

The amazing posture of this orchid easily identifies it from the preceding clubbed orchids; the upswept petals giving rise to its name. The other identifying feature is the calli arrangement in two double rows, instead of the usual four equally spaced rows.

The widespread distribution covers the lower south-west and extends north to Bindoon and east to Esperance.

It favours sandy to gravelly soils in Wandoo and Sheoak woodlands, and Jarrah forests.

My sites:	Map **12**, SW 22	Map **30**, SE 4
My best time:	late September, early October	

Caladenia macrostylis

Leaping Spider Orchid

Relative Caladenia Flower Size: small

The Leaping Spider Orchid is very distinctive. It has some similarity to the preceding orchids, but has only small knob-like clubs. The almost black calli are densely packed on the labellum apex.

The upswept petals produce a leaping appearance.

It is very widespread, although somewhat scattered, in a variety of habitats including Jarrah and Karri forests, coastal *Banksia* and Sheoak woodlands.

My sites: Map **5**, SW 7
Map **11**, SW 38
Map **12**, SW 22
Map **13**, SW 23, 26 b, 29
Map **14**, SW 32
Map **18**, SW 42
Map **42**, E 5

My best time: late September

Caladenia lobata

Butterfly Orchid

Relative Caladenia Flower Size: large

The Butterfly Orchid is one of the most distinctive and attractive of the *Caladenia* genus. The structural features of a broad, flat labellum, with long fringes, the upcurved lateral sepals, and the delightful yellow-green colouring make for an exciting find. The petals have a shape similar to the lateral sepals but without the clubs.

It grows on hill slopes near creeks and rivers between Bunbury and Cranbrook.

My sites: Map **21**, SC 2 Map **23**, SC 6 b

My best time: late October

Caladenia falcata

Fringed Mantis Orchid or, Green Spider Orchid

Relative Caladenia Flower Size: large

This orchid is found in a triangular area of the wheatbelt from Wongan Hills to Jerramungup to Albany. Its habitats vary – Wandoo and Jam woodlands, Sheoak thickets, and low shrublands, especially in wetter areas.

The structure is similar to the Butterfly Orchid, except that the lateral sepals have only one bend, with the remaining parts of the sepals being almost vertical.

The labellum has a smooth, narrow, deep red apex, and the flower is greener.

My sites: Map **26**, SC 11, 12
Map **28**, SE 1
Map **31**, SE 7
Map **32**, SE 9
Map **39**, E 2

My best time: early October

Caladenia attingens subsp. *attingens*

Forest Mantis Orchid

Relative Caladenia Flower Size: medium large

The Forest Mantis Orchid is located in the south-west between Harvey and Jerramungup, in the coastal and inland area to the west of the Fringed Mantis Orchid distribution. It has a variety of habitats – Jarrah, Marri, and Karri forests, and low heath. Soils vary also from deep sand to granitic and gravelly loams.

It is similar in appearance to the Fringed Mantis Orchid, but is generally more brightly coloured, has a more pointed labellum, and taller fringes. Also, the slender clubs are more prominent.

My sites: Map **5**, SW 6; Map **7**, SW 11, 12; Map **11**, SW 19; Map **12**, SW 22; Map **13**, SW 28, 29; Map **16**, SW 40; Map **19**, SW 43

My best time: early October

Caladenia attingens subsp.*gracillima*

Small Mantis Orchid

Relative Caladenia Flower Size: small

This orchid differs from the previous mantis orchids in being even smaller, and having a more easterly distribution from Jerramungup to Israelite Bay.

It favours Mallee and dense scrub, usually near winter-wet areas.

My site: Map **33**, SE 13

My best time: September

Caladenia extans

Pointing Spider Orchid

Relative Caladenia Flower Size: small

The Pointing Spider Orchid is restricted to small areas between Esperance and Israelite Bay. It is similar to the preceding mantis orchids, but has a more pointed and protruding labellum. The sepals have slender clubbing.

It is found in soil pockets on granite outcrops and sometimes in swampy Yate woodlands.

My site: Map **35**, SE 16

My best time: late September

Caladenia crebra

Arrowsmith Spider Orchid

Relative Caladenia Flower Size: small

This orchid is distinctive. It has a wide, flat labellum with a mass of dark calli. The fringes are prominent. The stiffly held, narrow petals are horizontal for more than half their length, and then curve downwards. Although the lateral sepals are clubbed, the dorsal sepal is not.

The almost iridescent green colouring of the orchid contrasts with the surrounding grasses and weeds.

It is often found near the trunks of Illyarrie trees (*Eucalyptus erythrocorys*) between Dongara and Jurien Bay.

My site: Map **49**, N 12

My best time: late August

Caladenia corynephora

Club-lipped Spider Orchid

Relative Caladenia Flower Size: small

It is not hard to see how this orchid justifies its name. The prominent clubbed tip of the labellum and very long fringes are unusual. The yellow-green sepals and petals have variable suffusions of maroon, and all have yellow-brown clubs.

It is restricted to small areas between Augusta and Albany where it favours winter-wet swamps.

My site: Map **13**, SW 25

My best time: late December

Caladenia roei

Ant Orchid or, Clown Orchid or, Jack-in-the-box or, Man Orchid

Relative Caladenia Flower Size: very small

This orchid is widespread throughout the wheatbelt and goldfields from north of Kalbarri almost to Esperance and across to Kojonup.

It has a fascinating structure, and when viewed from different angles, justifies the various common names by which it is known.

The very dark calli are concentrated at the back of the broad labellum that lacks the fringes of most *Caladenias*. The petals and sepals, which are usually yellow-green with maroon suffusions, are often curled at the ends.

My sites: Map **51**, N 14, 15, 17 Map **52**, N 18

My best time: early September

Caladenia multiclavia

Lazy Spider Orchid

Relative Caladenia Flower Size: very small

This is a peculiar-looking spider orchid. All the sepals and the petals are upcurved, with a central striped labellum "standing" on the bases of the sepals and petals. On the labellum there is a central band of calli ending at the apex with a mass of darker calli. The flowers are deep red and yellow.

It is found mainly in the southern wheatbelt, down to Ravensthorpe and Borden.

My site: Map **32**, SE 9

My best time: early October

Caladenia cairnsiana

Zebra Orchid

Relative Caladenia Flower Size: very small

The tiny Zebra Orchid is quite distinctive. Apart from its red-maroon striped labellum, the petals and dorsal sepals hang down close to the plant stem, with their tips slightly downcurved.

There is a mass of dark calli in the centre of the back section of the labellum, with a dark glandular tip at the apex.

The Zebra Orchid is widespread throughout the south-west, northwards to Lancelin and eastwards well past Esperance.

It is found in a variety of habitats, including open Mallee woodland, Peppermint woodland, Jarrah forest and coastal heath.

My sites: Map **3**, SW 4
Map **13**, SW 29
Map **19**, SW 44
Map **31**, SE 8
Map **32**, SE 9
Map **35**, SE 16
Map **42**, E 5

My best time: late September, early October

Caladenia discoidea

Bee Orchid or, Dancing Spider Orchid or, Antelope Orchid

Relative Caladenia Flower Size: very small

This is a distinctive orchid that cannot be confused with any other. Its scientific name is apt when considering the shape from above. As well as the short, curved petals and sepals it has very long fringes around the striped labellum. The plants usually carry several flowers.

It is found in a wide diagonal belt across the south-west from the west coast to the south coast, between Kalbarri and Busselton to Albany and Israelite Bay. It is not found in the wetter lower south-west. It generally grows in sandy *Banksia* woodland, but in its eastern habitat may be found in Mallee scrub and even salt lake margins.

My sites: Only Metropolitan sites, such as Kings Park, Wireless Hill, and Warwick Open Space.

My best time: September

Caladenia hirta subsp. *hirta*

Sugar Candy Orchid

Relative Caladenia Flower Size: small

The dainty Sugar Candy Orchid usually has several flowers with white floral parts, commonly tinged pale pink.

It is relatively common in the lower south-west and northwards as far as Arrowsmith. Its habitats vary from Tuart, Peppermint, *Banksia*, Sheoak, and Jarrah woodlands to the margins of winter-wet swamps. Soils vary from sand to gravelly loam.

My site:
Map **25**, SC 9

My best time:
October

Caladenia hirta subsp. *rosea*

Pink Candy Orchid

Relative Caladenia Flower Size: very small

The Pink Candy Orchid is similar to the Sugar Candy Orchid but has smaller flowers, a deeper pink colouring, shorter stems, usually occurs in clumps, and flowers earlier.

It is also found further east than the Sugar Candy Orchid, in the wheatbelt and goldfields from Kalbarri to Israelite Bay.

It has a variety of habitats including granite outcrops, salt lake margins and watercourses. Soils vary from sand to red-clay loams.

My sites: Map **26**, SC 11
Map **32**, SE 9
Map **33**, SE 13
Map **35**, SE 16
Map **51**, N 17
Map **52**, N 18, 19
Map **55**, N 22

My best time: late August, early September

Caladenia nana subsp. *nana*

Little Pink Fan Orchid

Relative Caladenia Flower Size: very small

This is another clump-forming orchid, often with many plants densely packed. The tiny flower ranges from pink to rose pink. The petals curve forwards, and the dorsal sepals are unusual in that they are joined near the base; they are also forward-facing.

It prefers well-drained soils in Jarrah forests. It is found in the lower south-west from Perth to Bremer Bay.

Note: *Caladenia nana* subsp. *unita*, or Pink Fan Orchid, is a very similar orchid but has more richly coloured, larger flowers, favours winter-wet swamps and flats, and flowers later.

My sites:	Map **11**, SW 21, 38	Map **18**, SW 42
My best time:	late September, early October	

Caladenia reptans subsp. *reptans*

Little Pink Fairy or, Dwarf Pink Fairy

Relative Caladenia Flower Size: very small

This is another clump-forming orchid of small stature – up to 15 cm. It is usually deep pink with the basal part of the labellum white or pale pink, while the apex is a very dark pink. The dorsal sepal is of similar shape to the petals and lateral sepals, while the width of the latter cause them to overlap at their bases.

It is very widespread, covering an area of the lower south-west from Kalbarri to Israelite Bay.

It favours Jarrah and Wandoo woodlands on the western side of the wheatbelt.

My site: Map **46**, N 4

My best time: early September

Caladenia latifolia

Pink Fairies

Relative Caladenia Flower Size: very small

This very common orchid is usually seen in attractive colonies near the bases of trees in Tuart and Peppermint woodlands, and also coastal heath. Soils are usually sandy, overlying limestone. Hence they are most often found in coastal areas between Kalbarri and Israelite Bay. They may also occur further inland near salt lakes and on granite outcrops.

The flowers vary from pink to deep pink, often with the tips of the floral parts almost white.

The inset shows an albino form.

My sites: Map **3**, SW 4 Map **16**, SW 40

My best time: late September

Caladenia marginata

White Fairy Orchid

Relative Caladenia Flower Size: very small

The White Fairy Orchid has a similar distribution to Pink Fairies, but prefers wet habitats around swamps and soil pockets on granite outcrops.

Its size and structure are very similar to the Pink Fairies, but the creamy-white flowers make them stand out when growing together.

My sites:	Map **13**, SW 26 b	Map **35**, SE 16
My best time:	late September	

Caladenia flava subsp. *flava*

Cowslip Orchid

Relative Caladenia Flower Size: very small

This extremely common orchid is widespread over the greater south-west and can be found almost anywhere, often in colonies of hundreds.

The flowers are so distinctive in colour and shape they need no description. Suffice to say that the red to brownish markings, and petal and sepal shape, vary from colony to colony. Sometimes the markings are completely absent (as illustrated).

My sites:
Map **3**, SW 4
Map **10**, SW 17
Map **11**, SW 20
Map **13**, SW 26 b, 28
Map **16**, SW 40
Map **20**, SC 1
Map **26**, SC 11
Map **27**, SC 13
Map **34**, SE 15
Map **44**, N 2
Map **48**, N 8

My best time: September and October

Caladenia flava subsp. *maculata*

Kalbarri Cowslip Orchid

Relative Caladenia Flower Size: very small

This orchid is restricted to an area from Perenjori, north through Kalbarri, to Nerren Nerren Station.

It generally has more blotchy markings, and bright lemon-yellow colouring to the floral parts. In most respects it is very similar to the common Cowslip Orchid.

My site: Map **50**, N 13

My best time: late August

Caladenia flava subsp. *sylvestris*

Karri Cowslip Orchid

Relative Caladenia Flower Size: very small

The Karri Cowslip Orchid is restricted to high rainfall areas of the lower south-west, especially Karri forest between Bunbury and Albany. It prefers dark, loamy soils and frequently inhabits the rotting trunks of fallen trees.

This orchid is usually taller than the other cowslip orchids and has paler flowers. It also has a later flowering period.

My site: Map **27**, SC 13

My best time: early November

Caladenia barbarossa

Common Dragon Orchid

Relative Caladenia Flower Size: very small

There are several dragon orchids in W.A. – this one is the most commonly found.

The petals and sepals are green and have deep maroon central bands. The floral parts are incurved with the tips often curled slightly. The labellum is quite exposed and covered in hairs. This orchid favours the western and southern wheatbelt in areas close to watercourses, especially in Sheoak and Wandoo woodlands. It has been found between Bindoon and Esperance.

The inset shows a lutea form.

My sites: Map **22**, SC 4
Map **25** SC 8
Map **31**, SE 8
Map **32**, SE 9

My best time: October

Praecoxanthus aphyllus

Leafless Orchid

Relative Caladenia Flower Size: very small

This is a pretty orchid which resembles the spider orchids but belongs to a different genus. The flower has a flattened appearance with the petals and sepals almost lying in a plane.

These floral parts are a bright cream-white and stand out against leaf litter. The labellum is prominent with attractive purple, green and yellow colouring.

It is found growing in sandy soil in *Banksia*, Sheoak and Jarrah country, from Pinjarra to Esperance.

My sites: Map **2**, SW 2 Map **9**, SW 14 b, 15

My best time: early May

Eriochilus dilatatus subsp. *dilatatus*

White Bunny Orchid

Flower length: 1¼-1¾ cm
Flower width: 1-1¼ cm

There are several bunny orchids to be found in W.A., nearly all in the wetter south-west. Most are only distinguishable by their leaves, as the flowers are so alike.

The White Bunny Orchid has long, narrow leaves. The flowers are much the same as others of the genus.

It prefers deep sand in coastal regions as part of Peppermint woodlands and Tuart forests. Occasionally it will be sighted on granite outcrops. The coastal strip it inhabits is between Shark Bay and Israelite Bay.

My sites:
Map **5**, SW 7
Map **9**, SW 14

My best time:
May

Eriochilus dilatatus subsp. *multifloris*

Common Bunny Orchid

Flower length: 1-1½ cm
Flower width: ¾-1¼ cm

The Common Bunny Orchid has small, oval-shaped leaves. The labellum is typically furry and spotted with maroon markings. The dorsal sepals are elongated and white, while the petals are reduced to shorter "ears".

The Common Bunny Orchid often has many flowers on each stem. It is mainly found in Jarrah forest and *Banksia* woodlands in the lower south-west, between Perth and Albany

My sites:

Map **2**, SW 2
Map **9**, SW 16
Map 11, SW 38
Map **15**, SW 35
Map **40**, E 3

My best time:

May

Eriochilus dilatatus subsp. *magnus*

Easter Bunny Orchid

Flower length: 1-2 cm
Flower width: 1-2 cm

This bunny orchid is best distinguished by its more widely spaced flowers when found with multiple flowers, and by its broader, oval leaf. The individual flowers are very similar to others of the genus.

It favours high rainfall areas of the lower south-west, between Harvey and Albany.

My sites:
Map **9**, SW 16
Map **15**, SW 35, 37

My best time:
May

Eriochilus dilatatus subsp. *undulatus*

Crinkle-leafed Bunny Orchid

Flower length: 1-1½ cm
Flower width: 1-1½ cm

This bunny orchid is distinguished best by its leaves. These are quite small but have a distinctive "crinkle" to their edges. The flowers are rarely multiple, but are structurally very similar to the preceding bunny orchids.

The distribution of this orchid also differentiates it because it favours the drier wheatbelt area, from Wongan Hills to the Fitzgerald National Park. Its habitats include salt lake margins, dense scrub, granite outcrops, and Mallee woodlands.

My site:
Map **40**, E 3

My best time:
May

Diuris corymbosa

Common Donkey Orchid or, Wallflower Orchid

Flower length:	2-3 cm
Flower width:	~2 cm

There are more than 20 species of donkey orchid in W.A., with this one being the most prolific. It can be found throughout the south-west and extending northwards to Dongara and eastwards to Esperance. It grows abundantly on the coastal plain, and especially in Jarrah forests.

As with other members of the genus, the petals form two large "donkey ears" which in this case are yellow, while the dorsal sepal is the central lobe behind the labellum. In this orchid these parts and the labellum wings have variable markings of brown and yellow.

My sites:
Map **7**, SW 12
Map **11**, SW 20
Map **13**, SW 26 b, 28, 29
Map **19**, SW 44
Map **25**, SC 9
Map **41**, E 4
Map **46**, N 4

My best time: September

Diuris aff. *corymbosa*

Sandplain Donkey Orchid

Flower length:	3-4 cm
Flower width:	2½-3 cm

This donkey orchid is similar to the Common Donkey Orchid but has a different coloration and is found in a quite restricted region. Although the literature suggests it is localised in the Bunbury area, I have found it near Gracetown, which is much further south.

It prefers *Banksia* woodlands, especially near winter-wet areas.

My sites: Map **4**, SW 5 Map **7**, SW 11, 12

My best time: late September

Diuris aff. *corymbosa*

Rosy-cheeked Donkey Orchid

Flower length: 3-5 cm
Flower width: 2-4 cm

This donkey orchid is a little larger than the preceding ones, but is similar in colour. The reddish-brown markings on the labellum lobes are the reason for the name. The lateral sepals are often curved and crossing over.

It is extensively distributed, mainly in the moist areas of the wheatbelt and goldfields, from Payne's Find to Morawa.

My sites: Map **51**, N 14, 17

My best time: early September

Diuris aff. *corymbosa*

Coastal Donkey Orchid

Flower length: 4-6 cm
Flower width: 3-4 cm

This beautiful orchid has been placed with *D. amplissima*, but is now regarded as a distinct species associated with *D. corymbosa*. It is found in the sandy soils of consolidated sand dunes between Augusta and Walpole.

My site: Map **16**, SW 40

My best time: October

Diuris pulchella

Beautiful Donkey Orchid

Flower length: 2-4 cm
Flower width: 2-3 cm

This donkey orchid is found over a wide area to the north-east of Esperance. It grows in shallow soils on granite outcrops.

It is identified by its location, its mauve-tinged labellum, and its side lobes.

My site: Map **35**, SE 16

My best time: early September

Diuris aff. *recurva*

Dainty Donkey Orchid

Flower length: ~2 cm
Flower width: ~2 cm

This small donkey orchid has distinctively rounder petals, which appear to be more widely separated than usual. The colouring is similar to that of the Common Donkey Orchid.

It inhabits damp areas of granite outcrops, creeks and especially open Wandoo woodland in an area from Harvey up to Mingenew and inland to Wubin and Dalwallinu.

My sites: Map **51**, N 16, 17 Map **54**, N 21

My best time: early September

Diuris aff. *amplissima*

Dunsborough Donkey Orchid

Flower length: 4-5 cm
Flower width: 2-3 cm

This beautiful donkey orchid has quite golden-brown petals and more deeply coloured sepals and lateral lobes. The labellum is also deeply coloured, usually with purple overtones. The lateral sepals are often curved and overlapping.

It inhabits coastal heath and Jarrah and Marri woodland in a strip between Dunsborough and Walpole.

My sites: Map **5**, SW 6, 7
Map **7**, SW 12
Map **16**, SW 40

My best time: early October

Diuris magnifica

Pansy Orchid

Flower length: 4-6 cm
Flower width: 3-5 cm

This stunning orchid has golden petals with variably toned brown and purple sepals and labellum lobes. The labellum has a distinctive fold down its centre. The Pansy Orchid is usually multiflowered and grows in colonies. Thus, it presents a colourful spectacle, as illustrated.

It favours limestone-based soils on the coastal plain, between Lancelin and Mandurah. It may also be found in *Banksia* and Sheoak woodlands.

My site: Map **51**, N 17

My best time: September

Diuris aff. *magnifica*

Arrowsmith Pansy Orchid

Flower length: 3-3½ cm
Flower width: 2½-3 cm

This orchid is closely related to the Pansy Orchid, but has smaller flowers, often with a more richly coloured labellum and lobes, and a more northerly distribution. Some varieties of this orchid show a lack of brown and a greater proportion of purple, as illustrated.

Its main habitats are sandy areas in light scrub and woodlands between Dongara and Regan's Ford.

My sites: Map **44**, N 2 Map **48**, N 6, 8

My best time: late August

Diuris longifolia

Purple Pansy Orchid

Flower length: 3-4 cm
Flower width: ~2 cm

This attractive, late-flowering orchid has definite mauve to purple colouring of the central floral parts, commonly with similar tingeing to the petals.

It grows in a range of habitats in the lower south-west, from Perth to Albany, favouring Karri and Jarrah forests, but also coastal heath and granite outcrops.

My sites: Map **5**, SW 7
Map **12**, SW 22
Map **13**, SW 26 b, 28
Map **16**, SW 40
Map **19**, SW 43
Map **22**, SC 4

My best time: October

Diuris heberlei

Heberle's Donkey Orchid

Flower length: 3½-4 cm
Flower width: ~3½ cm

This uncommon orchid is very attractive with its almost completely golden-yellow colouring, and sharply defined maroon-red markings at the base of the labellum and along the lateral sepals.

It is usually found in small, scattered groups in winter-wet flats in consolidated dunes near the coast, between Augusta and Albany. It has a very late flowering period. The orchid pictured here was photographed, with 20 or so others, in early February (2006).

My site: Map **14**, SW 34

My best time: late January

Diuris laxiflora

Bee Orchid

Flower length: 1-2 cm
Flower width: 1-2 cm

The Bee Orchid is very widespread, covering all the south-west below a line from Geraldton to Israelite Bay. Its habitats vary considerably but are generally winter-wet, and they are often found growing in profusion in several centimetres of water.

The bee orchids are smaller versions of the donkey orchids with similar colouring. This one has overall yellow with brown spots and markings.

Although widely distributed, the other bee orchids are fairly confined to small areas.

My sites: Map **5**, SW 7
Map **13**, SW 26 b
Map **26**, SC 11
Map **28**, SE 2
Map **32**, SE 9, 11
Map **33**, SE 13
Map **34**, SE 15
Map **35**, SE 16

My best time: early October

Diuris aff. *carinata*

Augusta Bee Orchid

Flower length: 1-2 cm
Flower width: 1-2 cm

This orchid is only known from the Augusta area and is found in relatively open areas.

The colouring is a little paler than the Bee Orchid, the markings similar, but the labellum is flatter and wider.

My site: Map **13**, SW 26 b

My best time: early November

Cyanicula sericea

Silky Blue Orchid

Flower length:	3-5 cm
Flower width:	3-4 cm

This is one of several of its genus with similar structure, and is characterised by the smooth petals and sepals that are all of similar size and shape. Although small, the flowers stand tall (up to 40 cm) and it is readily spotted in the bush. It often occurs as an albino form, as illustrated.

It favours gravelly soils in Jarrah forests and coastal *Banksia*.

My sites:	Map **11**, SW 18, 38	Map **16**, SW 40
	Map **13**, SW 23	Map **18**, SW 42

My best time: late September, early October

Cyanicula gemmata

Blue China Orchid

Flower length: 2-5 cm
Flower width: 2-5 cm

A very widespread blue orchid, the Blue China Orchid can be found almost anywhere in the south-west, extending past Kalbarri to the north, and past Esperance to the east. It grows best after summer fires, particularly in Jarrah forest, coastal heath, and low scrub.

The flowers are strikingly blue to mauve, and are carried on a short stem, usually no greater than 15 cm.

My sites: Map **3**, SW 4; Map **5**, SW 7; Map **6**, SW 9; Map **32**, SE 10, 12

My best time: early October

Cyanicula gertrudiae

Pale China Orchid

Flower length: 2-5 cm
Flower width: 2-5 cm

Similar to the Blue China Orchid, this orchid has paler blue flowers and a totally green leaf. The leaf of the Blue China Orchid is purple-brown underneath.

The Pale China Orchid is restricted to stabilised coastal dunes between Yallingup and Albany.

My sites: Map **5**, SW 6 Map **16**, SW 40

My best time: early October

Pheladenia deformis

Blue Fairy Orchid or, Blue Beard

Flower length: 2-3 cm
Flower width: 2-3 cm

The Blue Fairy Orchid is extremely widespread, encompassing the entire south-west below a line from well north of Kalbarri, directly towards Israelite Bay, and then easterly to the Great Australian Bight. The habitats are many and varied, including *Banksia* and Jarrah forests, coastal heath, swamp margins, granite outcrops and open flats that are winter-wet.

The flowers are blue to mauve with a caladenia-like structure, although the sepals and petals are of similar size and shape.

My sites: Map **44**, N 2
Map **46**, N 4
Map **47**, N 5
Map **48**, N 8
Map **51**, N 14, 16, 17
Map **52**, N 19
Map **55**, N 22

My best time: late August

Elythanthera brunonis

Purple Enamel Orchid

Flower length: 2-3 cm
Flower width: 1-3 cm

The Purple Enamel Orchid seems to be as commonly encountered as the Cowslip Orchid. Its distribution covers all the south-west below a line from Kalbarri to Israelite Bay. It is frequently found in spreading colonies among Wandoo and Jarrah forests, coastal heath, winter-wet flats and creek and river banks.

Although prolific, this orchid still retains its charm, especially when found in quantity. The glossy, purple flowers range in depth of colour, but are unmistakable.

My sites:
Map **5**, SW 7
Map **7**, SW 12
Map **11**, SW 39
Map **13**, SW 26 b, 28
Map **19**, SW 43
Map **23**, SC 6 a
Map **24**, SC 7
Map **25**, SC 10
Map **26**, SC 11
Map **32**, SE 10, 11, 12
Map **35**, SE 16
Map **37**, SE 19
Map **43**, N 1
Map **46**, N 4

My best time: late September, early October

Elythanthera emarginata

Pink Enamel Orchid

Flower length:	3-5 cm
Flower width:	3-5 cm

The Pink Enamel Orchid is generally much larger than the Purple Enamel Orchid, flowers later, has an obvious pink colour, and has a differently folded labellum. It is frequently found growing in colonies.

Although its distribution is as widespread as the Purple Enamel Orchid, it is far less profuse. It is found in the same habitats.

My sites: Map **13**, SW 23
Map **14**, SW 31, 33
Map **23**, SC 6 a
Map **24**, SC 7
Map **25**, SC 9
Map **26**, SC 11
Map **42**, E 5

My best time: late October, early November

Leptoceras menziesii

Rabbit Orchid

Flower length: 1-2 cm
Flower width: ~1 cm

The Rabbit Orchid is the only representative of the genus *Leptocerus*. It is characterised by the erect, deep red petals, forward-projecting, semi-oval white lateral sepals, and tiny size. It is usually found growing in large colonies.

It favours winter-wet areas throughout the south-west below a line from Kalbarri to Israelite Bay.

My sites: Map **1**, SW 1; Map **26**, SC 11; Map **35**, SE 16; Map **39**, E 2

My best time: late September

Leporella fimbriata

Hare Orchid

Flower length: 1-3 cm
Flower width: ~1 cm

The Hare Orchid is the only representative of the genus *Leporella*. Like the Rabbit Orchid, it has erect, ear-like petals, but has hanging lateral sepals and a large, smooth, green and brown labellum, with fringing but no calli. It generally grows in colonies, but because of its diminutive size and similar coloration to leaf litter, it can be exasperating to find.

It has a wide variety of habitats including sandy and gravelly areas in *Banksia* and Jarrah forest, coastal heath, scrub, and winter-wet areas near creeks and swamps.

The distribution includes all the south-west below a line from north of Kalbarri to Israelite Bay, then well east to the Great Australian Bight.

My sites:

Map **9**, SW 14 a
Map **38**, E 1
Map **45**, N 3

My best time:

May

Cyrtostylis huegelii

Midge Orchid

Flower length: 1-2 cm
Flower width: ~½ cm

This is one of three W.A. orchids of the genus *Cyrtostylis*, and is reported to be quite common. However, I have had difficulty in finding it and have not yet sighted the other two – the Mosquito Orchid and the Gnat Orchid.

These orchids all have similar structures with the first two being equally the largest. The Midge Orchid has brownish tonings, a narrower labellum and a larger distribution than the other two.

It is very widespread, particularly in moist areas and coastal areas of low heath. It can also be found in Jarrah and Marri forests and around granite outcrops further inland. It is usually found in large colonies of almost circular, ground-hugging leaves, few of which actually sprout flowers. The distribution is from Kalbarri to Esperance.

My sites:
Map **16**, SW 40
Map **30**, SE 5
Map **35**, SE 16

My best time:
late July

Lyperanthus serratus

Rattle Beaks

Flower length: ~2½ cm
Flower width: 3-5 cm

Rattle Beaks are the only member of the *Lyperanthus* genus in W.A.. There are invariably several flowers at the top of each stem. These cannot be mistaken for any other orchid, with their dull-green cap (dorsal sepal), mass of pale yellow calli, and protruding brown lined petals and lateral sepals. They are commonly found poking up through vegetation; especially Balga.

They are quite common throughout the wetter south-west and south coast as far as Israelite Bay. They favour several habitats, including Jarrah and Marri forests, coastal heath and woodlands.

My sites:
Map **7**, SW 11
Map **8**, SW 13
Map **10**, SW 17
Map **11**, SW 19, 39
Map **12**, SW 22
Map **13**, SW 29
Map **24**, SC 7
Map **28**, SE 1
Map **35**, SE 16

My best time: late September, early October

Pyrorchis nigricans

Red Beaks

Flower length: 2-3 cm
Flower width: 2-3 cm

Red Beaks are one of two species of *Pyrorchis* found in W.A., the other being Pink Beaks (*Pyrorchis forrestii*). Red Beaks is extremely widespread over the greater south-west from Shark Bay to Israelite Bay. Where the terrain is not wet enough, it inhabits moist areas around granite outcrops. The flowers are rarely seen in isolation, but mainly as large colonies in areas where summer bushfires have occurred. Otherwise the multitude of huge leaves (known as Elephants' Ears) cover the ground without producing flowers. The flowers of Red Beaks are similar to those of Rattle Beaks, but have quite different coloration and few calli.

My sites:
Map **8**, SW 13
Map **31**, SE 6
Map **32**, SE 10

My best time:
September

Drakaea livida

Warty Hammer Orchid

Flower length: 2-4 cm
Flower width: ⅓-½ cm

Of nine species of *Drakaea* in W.A., the Warty Hammer Orchid is fairly commonly found. It occurs throughout the south-west, from Watheroo to the Fitzgerald National Park. It usually favours well-drained soil in Jarrah forests.

Its common name is apt – the hammer shape being characteristic of the other members of the genus, and the distinctive dark maroon glands on the labellum looking like warts. The labellum is more hook-shaped than other hammer orchids.

The reduced petals and lateral sepals hang down around the stem, and the dorsal sepal is upcurved at the other end. The labellum hinge is a common feature of this orchid and the other hammer orchids.

Because of its diminutive size, it is difficult to find among leaf litter, but where old sand tracks are present, success is more likely.

My sites: Map **19**, SW 43, 44

My best time: late September

Drakaea species

Unidentified Hammer Orchid

Flower length: 1¼-2½ cm
Flower width: ¼-⅓ cm

This orchid is similar to the Dwarf Hammer Orchid, which is smaller than the other species of *Drakaea*. However, it differs in several ways and could not be identified by Andrew Brown. At this time it is regarded as an unnamed, rare species.

It grows in sandy *Banksia* woodlands on private property in the Dunsborough area.

My sites: Confidential

My best time: late September

Drakaea glyptodon

King-in-his-carriage

Flower length: 2-3 cm
Flower width: ⅓-½ cm

This hammer orchid is similar in structure to the other members of the genus, but is differently named. The labellum is characteristically shaped something like a horse's head. As with other orchids of the genus, this one is not easily found at first. However, if one persists in open sandy areas and old sand tracks, there are often large colonies to be discovered. I have found it most often in white sand, where the colours of the orchid contrast with the white background.

This orchid is widely scattered throughout the south-west, from Jurien Bay to Esperance. It favours sandy soils in various woodlands, and especially old sand tracks.

My sites: Map **8**, SW 13; Map **13**, SW 28; Map **19**, SW 43; Map **42**, E 5

My best time: late September, early October

Paracaleana nigrita

Flying Duck Orchid

Flower length:	1½-3 cm
Flower width:	1-1½ cm

This orchid is the most commonly found of about ten in the *Paracaleana* genus in W.A.. They have similar structures to one another and to the hammer orchids, except that the dorsal sepal hangs down like the petals, the labellum is on a curved 'neck' and can flip over onto insect pollinators, and the 'body' is the enlarged column wings, while the lateral sepals appear as 'wings'.

The Flying Duck Orchid is found in the lower south-west, from Bindoon to past Esperance. It favours sandy soils, and is easiest to find in old white sand tracks.

My sites: Map **8**, SW 13   Map **19**, SW 43

My best time: late September, early October

Thelymitra fuscolutea

Chestnut Sun Orchid

Flower length: ~3 cm
Flower width: ~3 cm

Thelymitras are often referred to as sun orchids because they only open fully in bright sunlight. There are over 30 species in W.A.. I have recorded 10 of these in this book.

This one, the Chestnut Sun Orchid, is typical in having the sepals and petals of similar shape and size; although the dorsal sepal is usually broader. Because the labellum of other orchids is a modified petal, the sun orchids lack a labellum, with three petals and three sepals.

The Chestnut Sun Orchid has a tufted appearance to the central column and the floral parts are mottled brown and dull gold.

It inhabits cooler forest areas over the lower south-west, between Perth and Albany. It prefers sandy soils overlying granite and laterite.

My sites: Map **13**, SW 26 b Map **14**, SW 33, 34

My best time: December

Thelymitra benthamiana

Leopard Orchid

Flower length:	~4 cm
Flower width:	~4 cm

This sun orchid is very widespread, from Northampton to Israelite Bay, and inland as far as Hyden. It is found in woodlands, winter-wet areas and in run-off areas from granite outcrops.

It is similar to the Chestnut Sun Orchid, but has lighter brown and more greenish yellow floral parts. It has a bulb-like projection on the column and rarely opens as fully as the Chestnut Sun Orchid.

My sites:	Map **12**, SW 22	Map **14**, SW 31
	Map **13**, SW 26 b	Map **32**, SE 12

My best time: early November

Thelymitra antennifera

Vanilla Orchid or, Lemon-scented Sun Orchid

Flower length: 2-3 cm
Flower width: 2-3 cm

This orchid is a very common one, and has one of the largest distributions of any orchid, from Shark Bay to the Great Australian Bight, and inland to Cue and the Goldfields. Like the preceding sun orchids, it prefers winter-wet areas and around granite outcrops.

The yellow colouring of this orchid is distinctive, as are the red-brown backs when the flowers are closed. The brown ear-like column lobes are also a feature.

My sites: Map **13**, SW 26 b; Map **25**, SC 9,10; Map **26**, SC 11; Map **28**, SE 2; Map **32**, SE 9; Map **35**, SE 16

My best time: late September, early October

Thelymitra variegata

Queen of Sheba

Flower length:	3-5 cm
Flower width:	3-5 cm

This orchid is a stunning sight when clusters are found in open areas with a drab background, as illustrated here.

The colours of the Queen of Sheba orchid vary a little, as do the pattern of markings, but the petals and sepals are fairly consistent within each flower. A small, corkscrew leaf is a feature of this orchid.

It is supposed to be found from Eneabba to Albany, in coastal plain sands. However, its sightings have dropped considerably in recent years and it may now be confined to the Eneabba-Gingin region.

My site: Map **48**, N 9

My best time: August

Thelymitra aff. *variegata*

Eastern Queen of Sheba

Flower length: 3-5 cm
Flower width: 3-5 cm

This orchid is similar to the preceding Queen of Sheba, with slightly deeper colours and more variability in the floral markings. Both forms, illustrated here, grow together.

The Eastern Queen of Sheba favours clay-loam and gravelly soils amongst low shrubs and extends from the Stirling Range to Condingup, east of Esperance.

My sites: Map **30**, SE 4 Map **31**, SE 6

My best time: late August

Thelymitra campanulata

Shirt Orchid or, Bell Sun Orchid

Flower length:	~2 cm
Flower width:	~2 cm

This orchid grows in near-coastal areas, from Kalbarri to Israelite Bay, omitting the stretch between Busselton and Albany. It tends to grow in sandy soils in *Banksia* woodland.

The Shirt Orchid has prominent striping on the petals and sepals, and the column has a distinctive yellow column-crest. The petals and sepals are more cup-like than those of other sun orchids, so the bell-shaped flower gives rise to its scientific name, and alternative common name.

My site: Map **43**, N 1

My best time: late September

Thelymitra crinita

Blue Lady Orchid or, Queen Orchid, or, Lily Orchid

Flower length: ~3 cm
Flower width: ~3 cm

This orchid has uniformly intense blue sepals and petals, with a prominent central column-crest of narrow black, orange and yellow calli. Each plant may have many flowers which provide a beautiful sight on a sunny day.

It appears to favour clay-loam soils in forest areas, and sandy soils in *Banksia* woodlands. The distribution is from Jurien Bay to Israelite Bay, and inland to the central wheatbelt.

My sites: Map **11**, SW 38
Map **12**, SW 22
Map **13**, SW 26 a
Map **18**, SW 42
Map **19**, SW 43

My best time: early November

Thelymitra aff. *macrophylla*

Granite Sun Orchid

Flower length: 1-2 cm
Flower width: 1-2 cm

The Granite Sun Orchid has a wide distribution throughout the wheatbelt and goldfields region, from Mt Magnet to Hyden and Israelite Bay.

It is normally found growing in small clumps in soil pockets on granite outcrops.

The flowers are the smallest of the sun orchids. The colour varies from pink to mauve, with a distinctive yellow apex to the column.

My site: Map **26**, SC 11

My best time: early October

Thelymitra aff. *macrophylla*

A Scented Sun Orchid

Flower length: 2-3 cm
Flower width: 2-3 cm

This is one of many scented sun orchids associated with the Scented Sun Orchid, *Thelymitra macrophylla*, but requiring further research to establish its proper taxonomy.

This orchid appears to be a late flowering variant occurring in high rainfall areas.

My site: Map **12**, SW 22

My best time: November

Thelymitra aff. *holmesii*

Plain Sun Orchid

Flower length:	2-3 cm
Flower width:	2-3 cm

The Plain Sun Orchid is much taller than most species of sun orchid, and usually has more flowers on its tall stem. These vary in colour from purple to pale blue, even to pink and white.

It occurs both as individual plants, and sometimes in large colonies near winter-wet swamps.

My sites: Map **12**, SW 22 Map **26**, SC 11

My best time: early November

Microtis alba

White Mignonette Orchid

Flower length: ½-1 cm
Flower width: ⅓-½ cm

The *Microtis* genus consists of commonly called mignonette, or onion, orchids. They have many tiny flowers on a spike, with a characteristic long, onion-like leaf.

The White Mignonette Orchid has one of the most compact flower spikes carrying up to 60 white flowers. The petals and sepals are reduced, but the labellum is relatively prominent. It has a frilly edge.

This orchid is distributed from Dongara to Israelite Bay in a coastal band extending over the lower south-west. It is found in a variety of habitats, including well-drained coastal heath, winter-wet swamp, woodlands, and Karri forests. It proliferates after summer fires.

My sites: Map **32**, SE 10, 11, 12

My best time: late October

Microtis media subsp. *media*

Common Mignonette Orchid

Flower length: ~⅓ cm
Flower width: ~⅓ cm

The Common Mignonette Orchid has up to 100 tiny white flowers and is similar to the White Mignonette, although the flowers have a less obtrusive labellum.

It is the most widespread of the mignonette orchids, extending from Shark Bay to the Great Australian Bight, and inland as far as Hyden. It has similar habitats to the White Mignonette Orchid.

My site:
Map **13**, SW 26 b

My best time:
October

Prasophyllum gracile

Little Laughing Leek Orchid

Flower length: ~1 cm
Flower width: ~½ cm

The genus *Prasophyllum* consists of a number of species commonly called leek orchids. There are more than 25 species in W.A.. All have reversed flowers, meaning that the labellum is above the column, rather than below, as in other orchids. In other respects they are similar to the mignonette orchids.

The Little Laughing Leek Orchid is characterised by having sharply-pointed lateral sepals forming an upward, barbed appearance.

This is a very common species covering a huge area, from Shark Bay to the Great Australian Bight, and inland to Wiluna and Laverton. It is mainly found in soil pockets on inland granite outcrops and in winter-wet areas in more coastal terrain.

My site: Map **26**, SC 11

My best time: September

Prasophyllum cyphochilum

Pouched Leek Orchid

Flower length: ~¾ cm
Flower width: ~¾ cm

This orchid is reasonably widespread, from Shark Bay to Israelite Bay, and inland to the central wheatbelt. It is mostly found in winter-wet areas, particularly Wandoo and Yate flats.

The orchid has pale yellow flowers that have a goblet shape. The pouched base of the labellum is the reason for the orchid's common name. The flowers are more widely spaced and there are up to 30 on each flower spike.

My sites:	Map **26**, SC 11	Map **36**, SE 17
My best time:	early October	

Prasophyllum parvifolium

Autumn Leek Orchid

Flower length: ~1 cm
Flower width: ~¾ cm

This attractive species has up to 20 flowers on the flower spike that prominently displays them. The larger flowers are attractive, with a large white wavy edged labellum, and sepals and petals with distinctly red-striped inner surfaces.

This orchid's habitat extends from Eneabba to Manjimup, and its habitats include Sheoak, *Banksia* and Jarrah woodlands, and coastal heath.

My site: Map **13**, SW 26 b

My best time: July

Prasophyllum aff. *parvifolium*

Scented Autumn Leek Orchid

Flower length: ~1 cm
Flower width: ~¾ cm

This leek orchid has up to 15 green and white flowers that hang outwards and downwards more than is usual.

It is commonly found in coastal heath, sandy *Banksia*, and Sheoak woodlands, and granite outcrops in Jarrah forests. The region extends from Bunbury to Israelite Bay.

My sites:
Map **9**, SW 14
Map **15**, SW 35, 36

My best time:
late May

Prasophyllum hians

Yawning Leek Orchid

Flower length: ~¾ cm
Flower width: ~¾ cm

This leek orchid is characterised by large, white petals and sepals, with frilly margins. There is a touch of pink at the base of the floral parts. It has up to 50 flowers on the tall flower spike which may be black or green.

It is found along a coastal strip from Dongara to Israelite Bay and extending over the lower south-west corner.

It flourishes after summer fires in a variety of habitats, including winter-wet flats and hillsides.

My site:
Map **51**, N 17

My best time:
September

Prasophyllum brownii

Christmas Leek Orchid

Flower length: 1-1½ cm
Flower width: 1-1¼ cm

The late-flowering Christmas Leek Orchid is one of the tallest species, often exceeding a metre in height. The stem carries up to 80 flowers. These are distinguished by the obvious stripes on the petals and sepals, contrasting with the white of the labellum. Like many others, the latter is frilly-edged.

It is usually located in the lower south-west, in the wetter Jarrah and Karri forests.

My sites:
Map **13**, SW 25, 27
Map **17**, SW 41

My best time:
late December

Pterostylis rogersii

Curled-tongue Shell Orchid

Flower length: 4-5 cm
Flower width: 1¼-1½ cm

The *Pterostylis* genus is represented in W.A. by over 70 species. I have included 14 in this book. In nearly all, the dorsal sepal and petals are united to form a hood. The labellum may be concealed, or protruding. The lateral sepals either stand upright forming 'ears', as in the shell and snail orchids, or are pointed downwards, as in the greenhoods. The *Pterostylis* have a rosette of leaves, either as a circle on the ground, or encircling the stem in a spiral, or both.

This shell orchid is a fascinating sight when a whole colony is in flower. The "head" is translucent and in bright sunlight the striped markings stand out.

It is found in coastal heath associated with Peppermint trees in a narrow belt of consolidated dunes, from Bunbury to Israelite Bay.

My sites: Map **15**, SW 35, 36

My best time: June

Pterostylis aff. *nana*

Limestone Snail Orchid

Flower length:	2½-3½ cm
Flower width:	½-¾ cm

As the common name suggests, this snail orchid is found on the coastal plain in sands overlying limestone. It appears to prefer the strip from Yanchep to Busselton. Its habitats include Tuart and *Banksia* woodlands, often associated with Zamias and Balgas.

This orchid usually grows in small colonies, has many small leaves, is relatively tall (up to 30 cm), and has quite long lateral sepals.

My site: Map **3**, SW 4

My best time: August

Pterostylis aff. *nana*

Slender Snail Orchid

Flower length: 3-4 cm
Flower width: ~½ cm

The Slender Snail Orchid is aptly named when its flower dimensions are compared to other snail orchids. Its stem is relatively short, has a small rosette of leaves on the ground, and the flowers have long lateral sepals.

It is usually found as clusters of several dozen plants, often concealed under low shrubs, and giving the appearance of sentinels on the alert.

They can be found in the Darling Range and coastal plain from Perth to Albany.

My sites: Map **3**, SW 4
Map **16**, SW 40
Map **18**, SW 42

My best time: early September

Pterostylis aff. *nana*

Short-eared Snail Orchid

Flower length: 1-1½ cm
Flower width: ~½ cm

This snail orchid is well named, as the lateral sepals are the shortest of its group. Its other features are similar to those of other snail orchids.

It is commonly found in coastal areas between Dongara and Israelite Bay, often forming colonies in the coastal heath and Peppermint and Tuart woodlands.

My site:
Map **3**, SW 4

My best time:
September

Pterostylis aff. *nana*

Hairy-stemmed Snail Orchid

Flower length: 2-2½ cm
Flower width: ½-¾ cm

The Hairy-stemmed Snail Orchid justifies its common name when compared with other snail orchids. Otherwise, the flower and leaves are very similar to the others. It is often found in colonies, particularly in wet conditions.

It is widespread throughout the wheatbelt and goldfields, mainly near salt lakes and granite outcrops.

My sites:
Map **51**, N 15, 16, 17
Map **53**, N 20

My best time:
late August

Pterostylis pyramidalis

Tall Snail Orchid

Flower length:	2-3 cm
Flower width:	½-¾ cm

This orchid is one of the tallest snail orchids, sometimes reaching over 30 cm. It is readily identified by its lack of a basal rosette of leaves, and having quite broad, short leaves – roughly heart-shaped, 1 to 2 cm long and ¾ to 1¼ cm wide.

It favours swampy areas and often forms large colonies.

It ranges down the coastal plain from Perth to Augusta.

My site:

Map **1**, SW 1

My best time:

August

Pterostylis spathulata

Spoon-lipped Rufous Greenhood

Flower length: 1¾-2½ cm
Flower width: 1-1½ cm

This greenhood is distinguishable by the pendulous tips to the sepals, and spoon-shaped labellum. The plant is relatively tall (up to 30 cm) and bears 2 to 10 flowers.

It is confined to the drier regions of the wheatbelt and goldfields from Mullewa to Ravensthorpe. It may be found around granite outcrops, in woodlands and in scrub.

My site:
Map **52**, N 18

My best time:
September

Pterostylis vittata

Banded Greenhood

Flower length: 1¼-1½ cm
Flower width: ¾-1 cm

This greenhood orchid has markedly striped flowers that are green and translucent. The flowers have cupped sepals, particularly as the they mature. There may be up to 25 flowers on a stem. It is relatively common in the lower south-west, but its distribution extends northwards to Dongara and eastwards to Israelite Bay where it is sparse.

It occurs in a variety of habitats from low scrub to Jarrah and Marri forests, and pine plantations.

My sites:

Map **1**, SW 1
Map **3**, SW 3
Map **11**, SW 20
Map **16**, SW 40
Map **26**, SC 11, 12

My best time:

August

Pterostylis sanguinea

Dark Banded Greenhood

Flower length: 1½-2½ cm
Flower width: 1¼-2 cm

This greenhood has variable colour, from green to brown, but is still distinctive. The example illustrated is a green variant. The flowers are readily identified by their broad, flat sepals.

It is widespread from Kalbarri to the Great Australian Bight, and inland to Hyden. However, it is not found in the lower south-west below a line from Perth to the Fitzgerald National Park.

The height of the plant varies considerably from 5 to 30 cm, and the flowers range in number from 2 to 10 or more. It favours areas around granite outcrops.

My sites:
Map **25**, SC 9
Map **31**, SE 6
Map **37**, SE 19
Map **45**, N 3

My best time:
August

Pterostylis aff. *sanguinea*

Coastal Banded Greenhood

Flower length: 1½-2 cm
Flower width: 1-1½ cm

The Coastal Banded Greenhood is similar to the other species of greenhood, but is distinguished by its narrow labellum and curved sepals. It generally grows in small groups, or singly, and each plant has up to 12 flowers.

It prefers sandy soils on the coastal plain, between Dongara and Bunbury, often relying on the protective shade of Zamia palms, Balgas, and scrub. Some outliers have been discovered in the gravelly soils of the Darling Range, and further south at Boyup Brook and Mt Barker.

My site:
Map **45**, N 3

My best time:
July

Pterostylis aff. *turfosa*

Late Bird Orchid

Flower length: 4-6 cm
Flower width: 1-1¼ cm

The Late Bird Orchid is so named because it flowers later than the other bird orchids. It has a densely tufted labellum and light network markings.

It can be found in Jarrah and Karri forests, in thickets adjacent to granite outcrops and in coastal Peppermint woodland. It is distributed between Busselton and Albany

My site: Map **13**, SW 28

My best time: late October, early November

Pterostylis barbata

Bird Orchid

Flower length:	4-5 cm
Flower width:	2-3 cm

The Bird Orchid is a delight to find. Its glossy coat, bird-like appearance, network markings, protruding sparsely-fringed labellum, and splayed lateral sepals are so distinctive. It usually grows singly, and with a height up to 30 cm, often appears above the weeds and leaf litter where it is growing.

It inhabits Jarrah forests and Sheoak woodlands from Perth to Albany.

My sites: Map **19**, SW 43, 44

My best time: late September

Pterostylis turfosa

Bearded Bird Orchid

Flower length: 5-7 cm
Flower width: ¾-1 cm

This orchid has a moderately hairy labellum and hanging lateral sepals. The "eye" is usually more evident than it is in any other of the bird orchids.

It is found as single plants in damp woodlands in south coastal areas.

My site: Map **22**, SC 4

My best time: early October

Pterostylis recurva

Jug Orchid

Flower length:	3-3½ cm
Flower width:	2½-3 cm

The Jug Orchid has a unique shape. It has little resemblance to other species of *Pterostylis*, except for the striped petals and sepals. The lateral sepals are extended and bent (hence the scientific name). It is a tall plant, often exceeding 50 cm in height and usually bearing single flowers, occasionally 2 or 3. The green and white flowers often show red and brown as they wither.

It is distributed widely throughout the south-west, from Geraldton to Israelite Bay, but is most common in the wetter, lower south-west forests.

My sites: Map **11**, SW 19, 38
Map **19**, SW 44
Map **24**, SC 7
Map **26**, SC 11
Map **36**, SE 17, 18
Map **37**, SE 19
Map **42**, E 5

My best time: September

Cryptostylis ovata

Slipper Orchid

Flower length: 2-3 cm
Flower width: ~¾ cm

This is the only representative of the *Cryptostylis* genus in W.A.. It is unique in many ways, most surprising is that it is the only W.A. orchid to have evergreen leaves. All other orchids' leaves wither and die each season. Its flower is reversed, with the dorsal sepal hanging below the flower, close to the stem. The lateral sepals point upwards, or curve upwards. The petals point backwards away from the large labellum, which is thrust forwards. The latter, pink-red to brown in colour, has network markings, reminiscent of the bird orchids.

The individual flowers on a stem appear one at a time, as the previous flower withers. There may be up to 15 of these, so flowering extends over a long period.

It can be found throughout the south-west from Perth to Fitzgerald National Park, but favours the wetter south coast, particularly in woodlands and forests.

My site: Map **14**, SW 30

My best time: December, January

Hybrids

There are many instances of orchids cross-pollinating with other orchids of different species in the same locality. The resulting flowers often show some characteristics of both parents and sometimes a blending of features. This occurs particularly amongst the *Caladenias*. This can be confusing. A particular discovery may be one of the species not included in this book, or it may be a hybrid. If it is a hybrid there will usually be clues about either one parent, or both.

Here are some of my 'finds' for you to consider:

Glossary

Adaptation a change of structure or behaviour, that promotes the likelihood of a plant's survival e.g. insect-attracting colours, perfume, labellum shape, etc.

Aff. literally, affinity; meaning the species is very closely related to another of the same genus, usually not yet named officially.

Albino abnormal flower form lacking pigmentation, appearing white.

Apex/Apical top, or tip, especially of the labellum of orchids.

Base/Basal bottom end of flower parts, attached to the stem.

Callus/Calli small projections on the labellum, usually in rows. Also, the fringes may be referred to as marginal calli.

Clubs/Clubbing a swelling towards the ends of sepals and petals, usually brownish in colour, and consisting of many tiny glandular cells.

Column a fused combination of the male stamens and female pistil of a flower. Usually above the labellum.

Confidential I have used this term to avoid revealing specific sites for orchids currently declared as Rare Flora. Many of these habitats are fragile, and the orchids are liable to extinction if protective measures are not maintained.

Dorsal the back, or upper, part of an organism or structure.

Endemic native to a particular place.

Epiphyte a plant which grows on another plant, but is not parasitic e.g. tropical orchids living in trees.

Fringe/Fringing a row of calli arising from the margins of the labellum in most Caladenia species. They vary in length, shape, and colour between species.

Genus/Genera a group of organisms with closely related characteristics. Some members can interbreed.

Habitat the specific environment in which a plant lives e.g. soil pockets on granite outcrops.

Hybrid the offspring of cross-fertilisation by parents of different species. Hybrid orchid flowers often show characteristics of both parents.

Labellum a modified lower petal, particularly common in orchids, forming a 'lip'.

Lateral from the side of a structure.

Lutea abnormal form of flower lacking red colouring; expressed as yellow.

Mallee a large number of Eucalypt species which develops numerous slender stems from an underground root stock, instead of a single trunk; usually enhanced by bush-fires.

Margin/Marginal referring to the edge, usually of a structure such as leaf, labellum, etc.

Petal a segment of the interior ring of floral parts. With orchids, there are 3, although one is usually modified to form a labellum, leaving 2 normal ones, placed laterally.

Pollination the transfer of (male) pollen from the anthers to the (female) stigma. This is usually followed by fertilisation, when the pollen grows down to unite with the ovules in the ovary.

Rare Flora several species in the book are declared as Rare Flora."These are taxa which have been adequately searched for and are deemed to be in the wild either rare, in danger of extinction, or otherwise in need of special protection, and have been gazetted as such".

Relative Caladenia Flower Size
an arbitrary indication of flower size devised by me to separate species on an approximate basis:

Classification	Flower Length (cm)	Flower Width (cm)	Spider Orchid Examples
very large	12 – 30	8 – 22	Noble, Splendid
large	7 – 18	5 – 12	Chapman's, Common
small	3 – 6	3 – 8	Exotic, Rusty
very small	2 – 4	2 – 4	Zebra, Winter

Scape the flower-bearing stem of a plant.

Sepal a segment of the exterior ring of floral parts. With orchids, there are normally 3, often shaped and coloured like the petals.

Site a particular location in which an orchid grows, or has been observed to grow.

Species closely related organisms with common characteristics which allow them to reproduce among themselves, but not commonly with other species. Usually designated by two Latin names in italics e.g. Caladenia arenicola.

Suffusions colouring pigments in the floral parts of orchids which may appear as stripes, dots, blotches, etc.

Taxon/Taxa a general term for any category of classification such as Family, Genus, Phylum, Species, etc.

Terrestrial describes organisms which have their habitats on, or in, the ground e.g. all south-west orchids. c.f. arboreal, aquatic, etc.

~ symbol for approximate e.g. ~5 cm means approximately 5 cm.

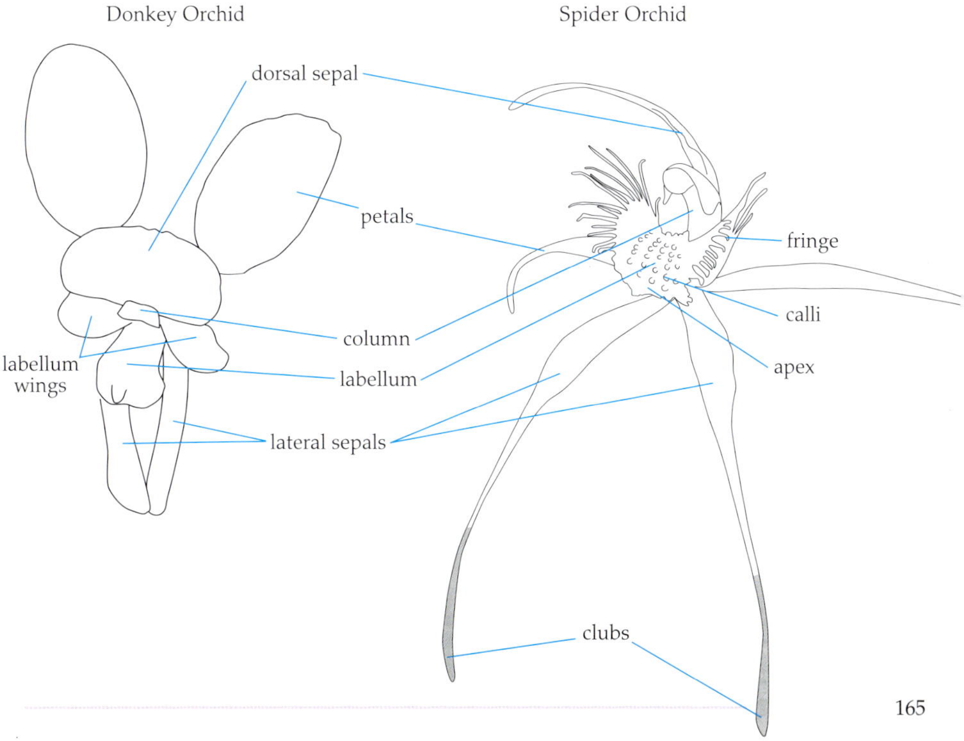

Mud Maps

MAP 1

SW 1: Paganoni Road,
off Mandurah Road,
near Golden Bay

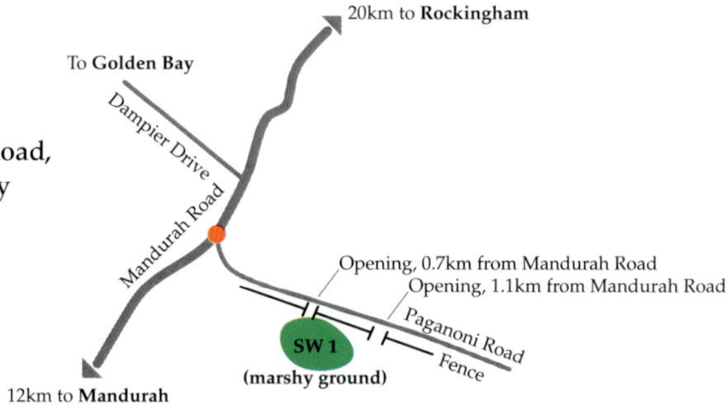

SW 1
Banded Greenhood (Aug)
Rabbit Orchid (Sep)
Tall Snail Orchid (Aug)

MAP 2

SW 2: Birchmont Road,
off Old
Bunbury Road

SW 2
Common Bunny Orchid (May)
Leafless Orchid (May)

MAP 3

SW 3: Preston Beach Road,
off Old Coast Road
SW 4: Parking, Old Coast Road,
6km south of Preston Beach Road

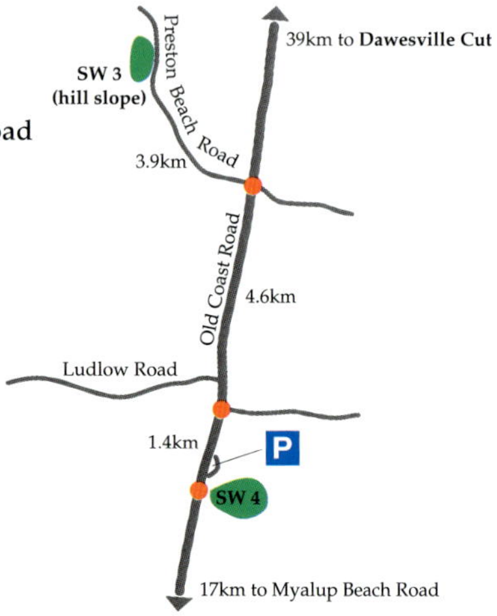

SW 3
Banded Greenhood (Aug)
Common Spider Orchid (Aug)

SW 4
Blue China Orchid (Oct)
Common Spider Orchid (Aug)
Cowslip Orchid (Sep)
Limestone Snail Orchid (Aug)
Noble Spider Orchid (Sep)
Pink Fairies (Sep)
Short-eared Snail Orchid (Sep)
Slender Snail Orchid (Sep)
Tuart Spider Orchid (Sep)
Zebra Orchid (Sep)

MAP 4

SW 5: Parking area, Bussell Highway, 11km south of Bunbury Ring Road

SW 5

Sandplain Donkey Orchid (Sep)
White Spider Orchid (Oct)

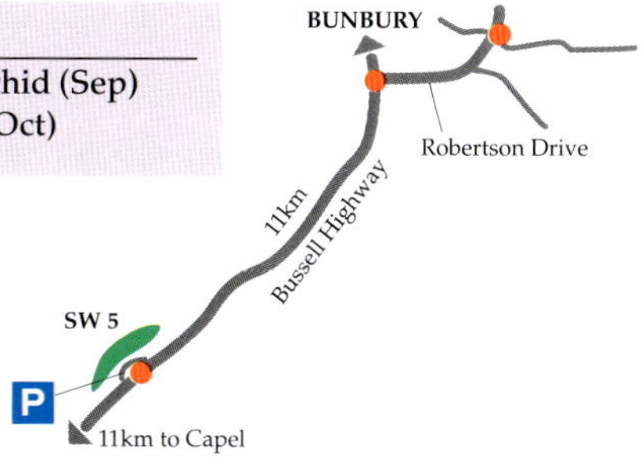

MAP 5

SW 6: Blythe Reserve, Dunsborough
SW 7: Meelup Regional Park walk trail

SW 6

Chapman's Spider Orchid (Oct)
Dunsborough Donkey Orchid (Oct)
Forest Mantis Orchid (Oct)
Karri Spider Orchid (Oct)
Pale China Orchid (Oct)
Rusty Spider Orchid (Sep)
Swamp Spider Orchid (Oct)

SW 7

Bee Orchid (*Diuris*) (Oct)
Blue China Orchid (Oct)
Chapman's Spider Orchid (Oct)
Dunsborough Donkey Orchid (Oct)
Hill's White Spider Orchid (Oct)
Leaping Spider Orchid (Sep)
Purple Enamel Orchid (Sep)
Purple Pansy Orchid (Oct)
Rusty Spider Orchid (Sep)
Swamp Spider Orchid (Oct)
White Bunny Orchid (May)

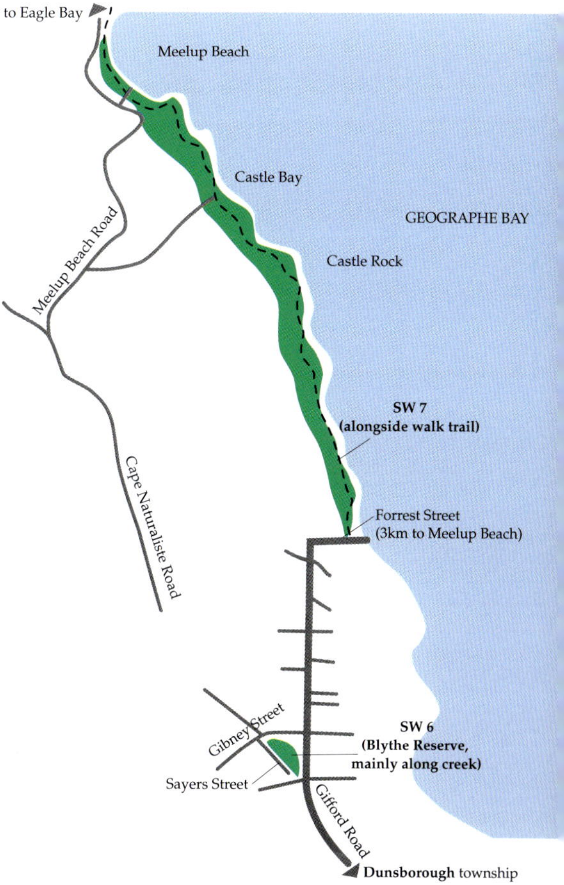

MAP 6

SW 8: Cape to Cape walk trail, Yallingup, cliff top car park
SW 9: Caves Road, 100m south of Old Yallingup Hall
SW 10: Caves Road, 1.5km south of Old Yallingup Hall

SW 8
Exotic Spider Orchid (Sep)

SW 9
Blue China Orchid (Oct)
Cherry Spider Orchid (Oct)
Rusty Spider Orchid (Sep)
Scott River Spider Orchid (Oct)

SW 10
Chapman's Spider Orchid (Oct)
Rusty Spider Orchid (Sep)
Scott River Spider Orchid (Oct)
Swamp Spider Orchid (Oct)

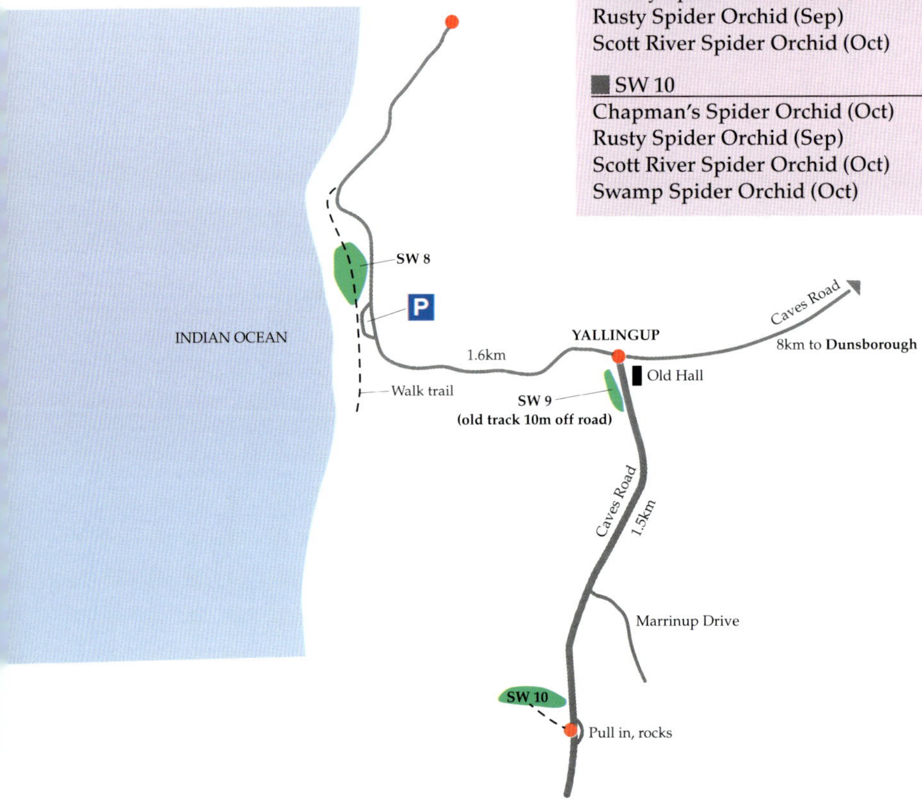

MAP 7

SW 11: Gracetown sand tracks
SW 12: Gracetown, hill

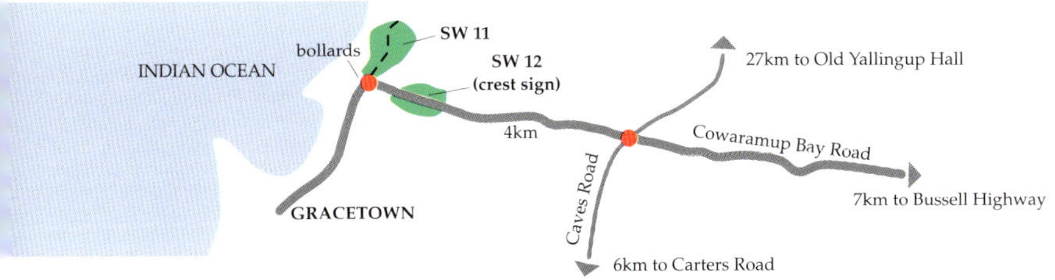

SW 11
Broad-lipped Spider Orchid (Sep)
Forest Mantis Orchid (Oct)
Funnel-web Spider Orchid (Oct)
Rattle Beaks (Sep)
Sandplain Donkey Orchid (Sep)

SW 12
Common Donkey Orchid (Sep)
Dunsborough Donkey Orchid (Oct)
Forest Mantis Orchid (Oct)
Purple Enamel Orchid (Sep)
Sandplain Donkey Orchid (Sep)
Scott River Spider Orchid (Oct)
Swamp Spider Orchid (Oct)

MAP 8

SW 13: Leeuwin Naturaliste National Park, WNW of Margaret River

SW 13
Flying Duck Orchid (Sep)
King-in-his-carriage (Sep)
Rattle Beaks (Sep)
Red Beaks (Sep)

MAP 9

SW 14 a, b: Caves Road, 2km south of Redgate Road
SW 15: Forest Grove Road, off Caves Road
SW 16: Caves Road, 3.5km south of Forest Grove Road

SW 14
Hare Orchid (May)
Leafless Orchid (May)
Scented Autumn Leek Orchid (May)
White Bunny Orchid (May)

SW 15
Leafless Orchid (May)

SW 16
Common Bunny Orchid (May)
Easter Bunny Orchid (May)

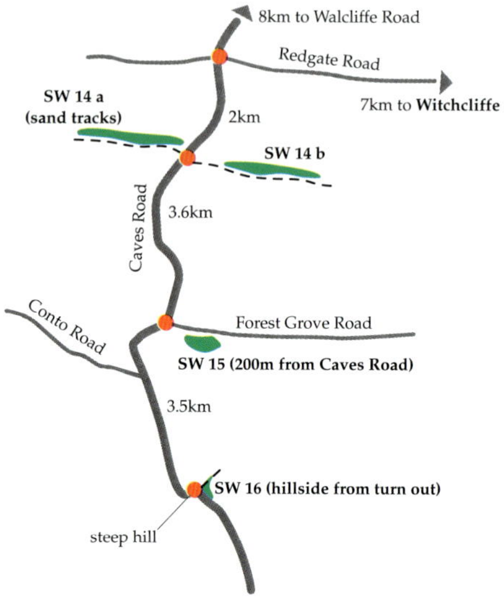

MAP 10

SW 17: Bussell Highway, Carbunup

SW 17
Chapman's Spider Orchid (Oct)
Cowslip Orchid (Sep)
Rattle Beaks (Sep)
Rusty Spider Orchid (Sep)

MAP 11

SW 18: Rowe Road, Witchcliffe
SW 19: Bussell Highway, south of Witchcliffe
SW 20: Bullant Drive, 9km south of Witchcliffe
SW 21: McLeod Creek Road Parking Bay
SW 38: Gravel Pit, Alexandra Bridge
SW 39: Warner Glen Road verge

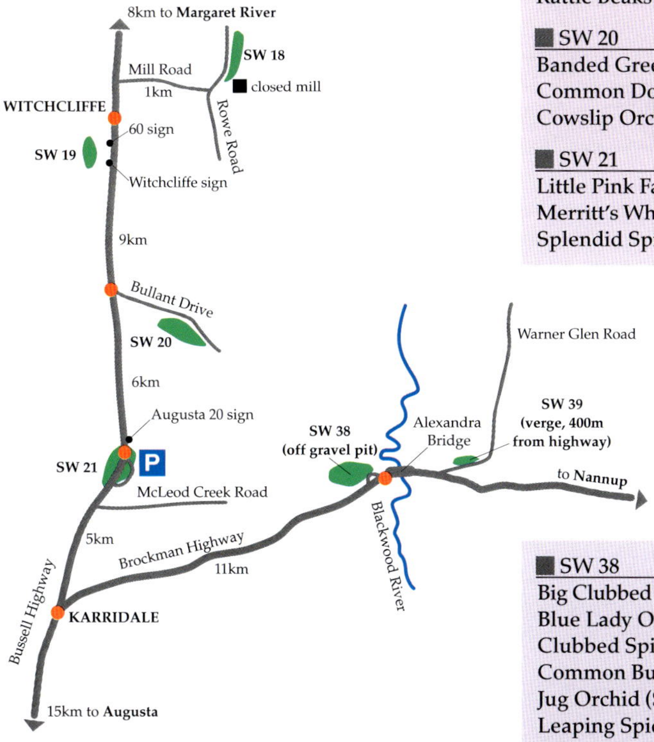

SW 18

Rusty Spider Orchid (Sep)
Silky Blue Orchid (Sep)

SW 19

Forest Mantis Orchid (Oct)
Jug Orchid (Sep)
Margaret River Spider Orchid (Sep)
Rattle Beaks (Sep)

SW 20

Banded Greenhood (Aug)
Common Donkey Orchid (Sep)
Cowslip Orchid (Sep)

SW 21

Little Pink Fan Orchid (Sep)
Merritt's White Spider Orchid (Oct)
Splendid Spider Orchid (Oct)

SW 38

Big Clubbed Spider Orchid (Oct)
Blue Lady Orchid (Nov)
Clubbed Spider Orchid (Oct)
Common Bunny Orchid (May)
Jug Orchid (Sep)
Leaping Spider Orchid (Sep)
Little Pink Fan Orchid (Sep)
Merritt's White Spider Orchid (Oct)
Silky Blue Orchid (Sep)

SW 39

Merritt's White Spider Orchid (Oct)
Purple Enamel Orchid (Sep)
Rattle Beaks (Sep)
Splendid Spider Orchid (Oct)

MAP 12

SW 22: Kudardup, behind Transfer Station

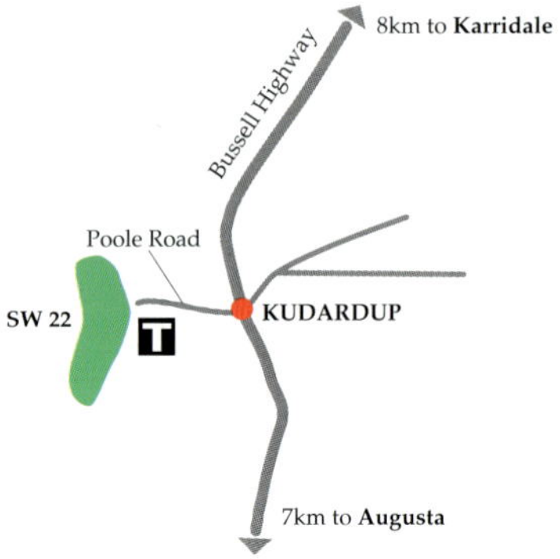

SW 22

- Blue Lady Orchid (Nov)
- Forest Mantis Orchid (Oct)
- Funnel-web Spider Orchid (Oct)
- Karri Spider Orchid (Oct)
- Leaping Spider Orchid (Sep)
- Leopard Orchid (Nov)
- Plain Sun Orchid (Nov)
- Purple Pansy Orchid (Oct)
- Rattle Beaks (Sep)
- Reaching Spider Orchid (Sep)
- Scented Sun Orchid, A (Nov)

MAP 13

SW 23: Augusta Yacht Club entrance road
SW 24: East-side S-bend, Yacht Club road
SW 25: Opposite side track, west-side, S-bend
SW 26 a, b: Flat Rocks, West Bay, Augusta
SW 27: Back track, Flat Rocks to Bussell Highway
SW 28: Augusta Information Bay
SW 29: West Bay Creek Road, Augusta

SW 23

Funnel-web Spider Orchid (Oct)
Leaping Spider Orchid (Sep)
Pink Enamel Orchid (Oct)
Silky Blue Orchid (Sep)

SW 24

Albany Spider Orchid (Dec)

SW 25

Albany Spider Orchid (Dec)
Christmas Leek Orchid (Dec)
Club-lipped Spider Orchid (Dec)

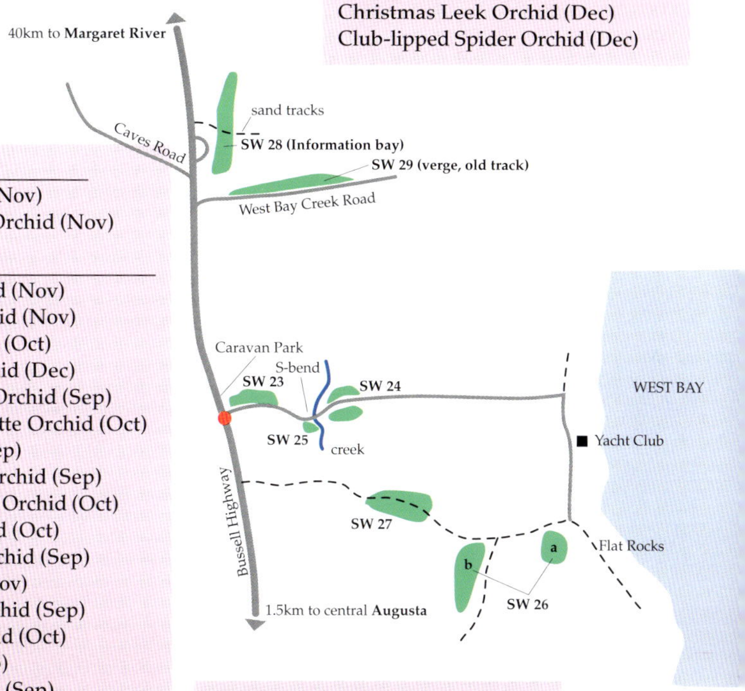

SW 26 a)

Blue Lady Orchid (Nov)
Christmas Spider Orchid (Nov)

SW 26 b)

Augusta Bee Orchid (Nov)
Autumn Leek Orchid (Nov)
Bee Orchid (*Diuris*) (Oct)
Chestnut Sun Orchid (Dec)
Common Donkey Orchid (Sep)
Common Mignonette Orchid (Oct)
Cowslip Orchid (Sep)
Diamond Spider Orchid (Sep)
Funnel-web Spider Orchid (Oct)
Karri Spider Orchid (Oct)
Leaping Spider Orchid (Sep)
Leopard Orchid (Nov)
Purple Enamel Orchid (Sep)
Purple Pansy Orchid (Oct)
Vanilla Orchid (Sep)
White Fairy Orchid (Sep)

SW 27

Albany Spider Orchid (Dec)
Christmas Leek Orchid (Dec)

SW 28

Common Donkey Orchid (Sep)
Cowslip Orchid (Sep)
King-in-his-carriage (Sep)
Forest Mantis Orchid (Oct)
Late Bird Orchid (Oct)

SW 28 (continued)

Purple Enamel Orchid (Sep)
Purple Pansy Orchid (Oct)
Rusty Spider Orchid (Sep)

SW 29

Common Donkey Orchid (Sep)
Forest Mantis Orchid (Oct)
Leaping Spider Orchid (Sep)
Rattle Beaks (Sep)
Zebra Orchid (Sep)

MAP 14

SW 30: Par 3 Golf Course, 5th tee, Augusta
SW 31: Augusta swamp
SW 32: Walk trail, old railway line, Augusta
SW 33: Road to old dump (closed), Augusta
SW 34: Swamp, off Skippy Rock Road, Flinders Bay

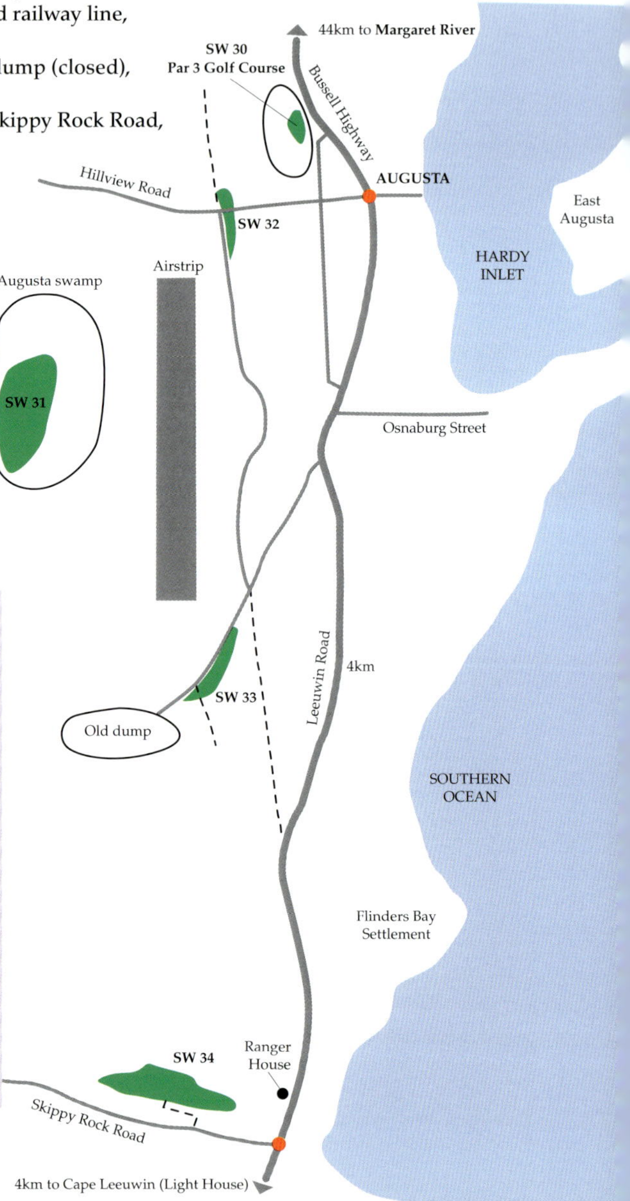

SW 30
Slipper Orchid (Dec)

SW 31
Leopard Orchid (Nov)
Pink Enamel Orchid (Oct)

SW 32
Clubbed Spider Orchid (Oct)
Funnel-web Spider Orchid (Oct)
Leaping Spider Orchid (Sep)
Merritt's White Spider Orchid (Oct)

SW33
Chestnut Sun Orchid (Dec)
Pink Enamel Orchid (Oct)

SW 34
Chestnut Sun Orchid (Dec)
Heberle's Donkey Orchid (Jan)

MAP 15

SW 35: Sand track off bend in Hillview Road, Augusta
SW 36: Bush off Green Hill Road, Augusta
SW 37: Augusta Golf Course

SW 35
Common Bunny Orchid (May)
Curled-tongue Shell Orchid (Jun)
Easter Bunny Orchid (May)
Scented Autumn Leek Orchid (May)

SW 36
Curled-tongue Shell Orchid (Jun)
Scented Autumn Leek Orchid (May)

SW 37
Easter Bunny Orchid (May)

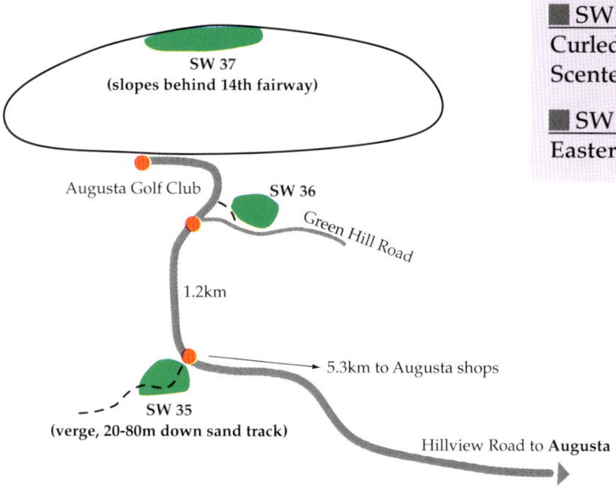

MAP 16

SW 40: Scott National Park

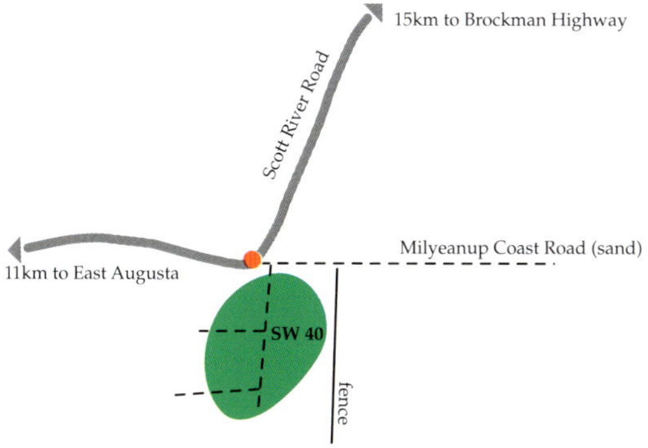

SW 40
Banded Greenhood (Aug)
Coastal Donkey Orchid (Oct)
Cowslip Orchid (Sep)
Dunsborough Donkey Orchid (Oct)
Forest Mantis Orchid (Oct)
Merritt's White Spider Orchid (Oct)
Midge Orchid (Jul)
Pale China Orchid (Oct)
Pink Fairies (Sep)
Purple Pansy Orchid (Oct)
Scott River Spider Orchid (Oct)
Silky Blue Orchid (Sep)
Slender Snail Orchid (Sep)
Stumpy Spider Orchid (Oct)

MAP 17

SW 41: Sues Road, south end off Brockman Highway

SW 41
Christmas Leek Orchid (Dec)

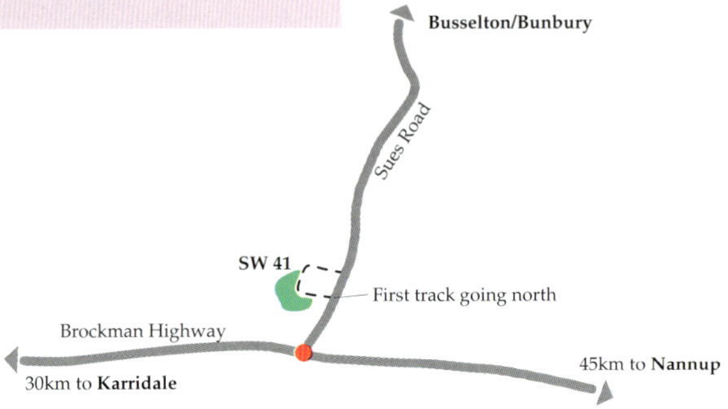

MAP 18

SW 42: Stewart Road, off Brockman Highway

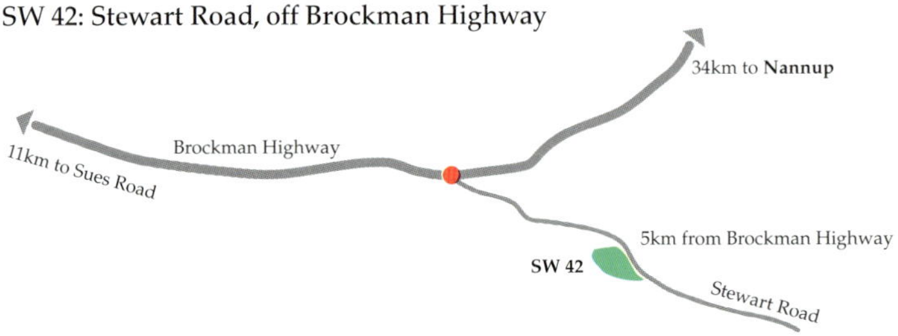

SW 42
Blue Lady Orchid (Nov)
Leaping Spider Orchid (Sep)
Little Pink Fan Orchid (Sep)
Silky Blue Orchid (Sep)
Slender Snail Orchid (Sep)

MAP 19

SW 43: Nannup, Kondil Park walk trails
SW 44: Nannup, Old Cundinup Road, off Vasse Highway

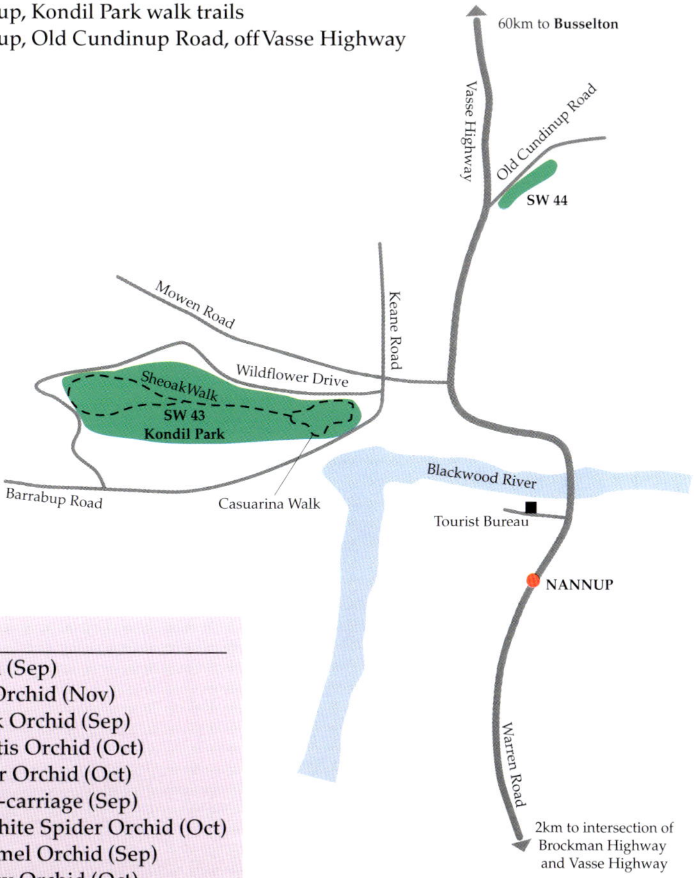

SW 43

Bird Orchid (Sep)
Blue Lady Orchid (Nov)
Flying Duck Orchid (Sep)
Forest Mantis Orchid (Oct)
Karri Spider Orchid (Oct)
King-in-his-carriage (Sep)
Merritt's White Spider Orchid (Oct)
Purple Enamel Orchid (Sep)
Purple Pansy Orchid (Oct)
Splendid Spider Orchid (Oct)
Warty Hammer Orchid (Sep)

SW44

Bird Orchid (Sep)
Common Donkey Orchid (Sep)
Jug Orchid (Sep)
Merritt's White Spider Orchid (Oct)
Warty Hammer Orchid (Sep)
Zebra Orchid (Sep)

MAP 20

SC 1: Falls Road, Serpentine, off South Western Highway

SC 1
Cowslip Orchid (Sep)

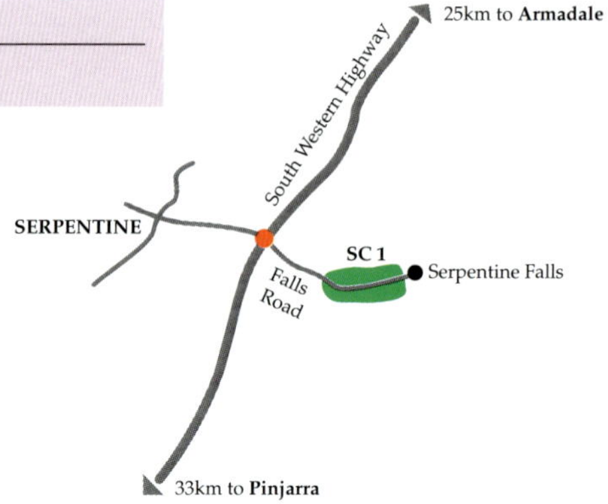

MAP 21

SC 2: Westbourne Road, 27km south of Boyup Brook, off Boyup Brook–Cranbrook Road

SC 2
Butterfly Orchid (Oct)
Mustard Orchid (Oct)

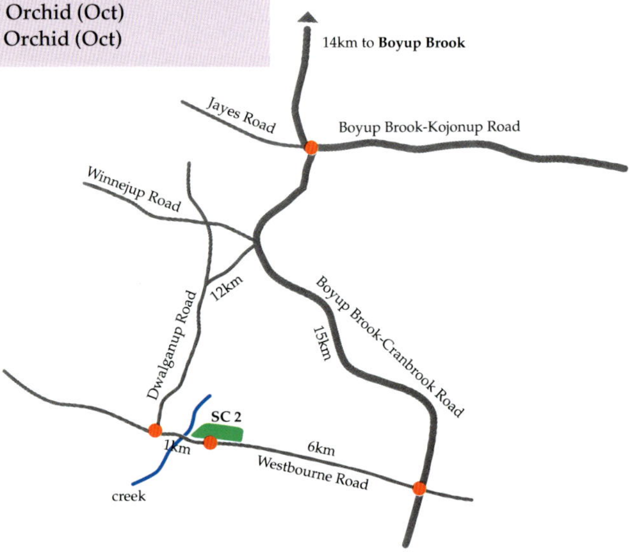

MAP 22

SC 3: Tone River, north of Muirs Highway, off Burangurup Road
SC 4: Lake Muir, east of Lookout
SC 5: 10km east of Red Lake Road

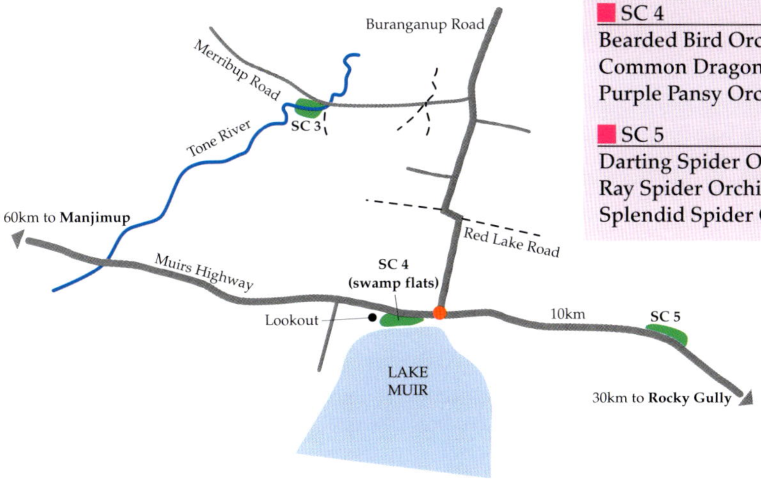

SC 3
Splendid Spider Orchid (Oct)

SC 4
Bearded Bird Orchid (Oct)
Common Dragon Orchid (Oct)
Purple Pansy Orchid (Oct)

SC 5
Darting Spider Orchid (Oct)
Ray Spider Orchid (Nov)
Splendid Spider Orchid (Oct)

MAP 23

SC 6 a,b: Frankland River–Muirs Highway bridge

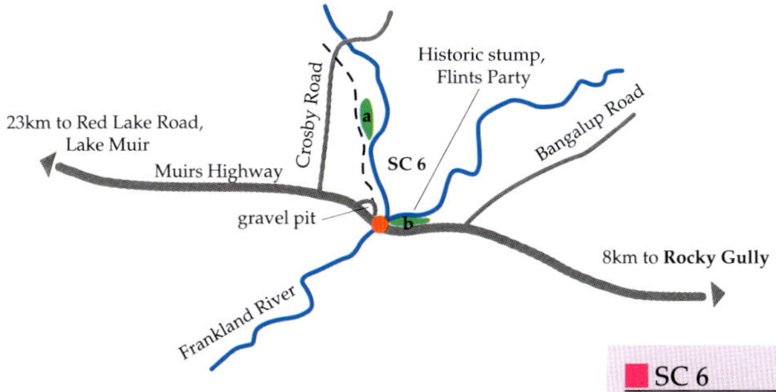

SC 6
Butterfly Orchid (Oct)
Karri Spider Orchid (Oct)
King Spider Orchid (Oct)
Pink Enamel Orchid (Oct)
Purple Enamel Orchid (Sep)

MAP 24

SC 7: Gravel pit, Woogenilup Road, off Albany Highway, Mt Barker

SC 7
Jug Orchid (Sep)
Pink Enamel Orchid (Oct)
Purple Enamel Orchid (Sep)
Rattle Beaks (Sep)
Stark White Spider Orchid (Oct)

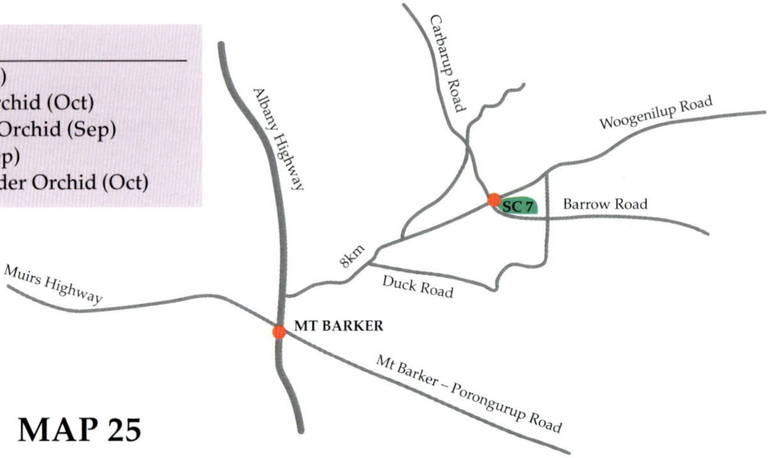

MAP 25

SC 8: Tenterden, intersection Archers Road, Lunt Road
SC 9: Cranbrook, wildflower walk
SC 10: Cranbrook, sand track

SC 8
Common Dragon Orchid (Oct)
Joseph's Spider Orchid (Oct)
Mustard Orchid (Oct)

SC 9
Common Donkey Orchid (Sep)
Dark Banded Greenhood (Aug)
Joseph's Spider Orchid (Oct)
Pink Enamel Orchid (Oct)
Stark White Spider Orchid (Oct)
Sugar Candy Orchid (Oct)
Vanilla Orchid (Sep)

SC 10
Joseph's Spider Orchid (Oct)
Purple Enamel Orchid (Sep)
Stark White Spider Orchid (Oct)
Vanilla Orchid (Sep)

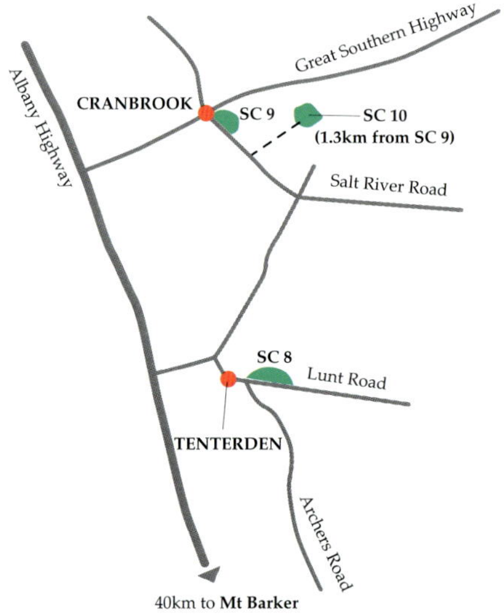

MAP 26

SC 11: Birdwood Nature Reserve, Mission Road, Kojonup
SC 12: Robinson Road, gravel dump

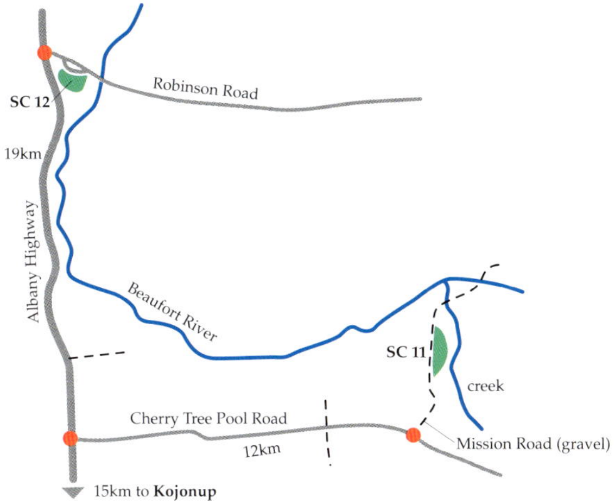

SC 11

Banded Greenhood (Aug)
Bee Orchid (*Diuris*) (Oct)
Common Spider Orchid (Aug)
Cowslip Orchid (Sep)
Crimson Spider Orchid (Sep)
Fringed Mantis Orchid (Oct)
Granite Sun Orchid (Oct)
Joseph's Spider Orchid (Oct)
Jug Orchid (Sep)
Little Laughing Leek Orchid (Sep)
Pink Candy Orchid (Aug)
Pink Enamel Orchid (Oct)
Plain Sun Orchid (Nov)
Pouched Leek Orchid (Oct)
Purple Enamel Orchid (Sep)
Rabbit Orchid (Sep)
Stark White Spider Orchid (Oct)
Vanilla Orchid (Sep)

SC 12

Banded Greenhood (Aug)
Fringed Mantis Orchid (Oct)

MAP 27

SC 13: Rest bay, Muirs Highway, 20km east of Manjimup

SC 13

Cowslip Orchid (Sep)
Karri Cowslip Orchid (Nov)

MAP 28

SE 1: Old gravel pit, Katanning, Katanning–Nyabing Road
SE 2: Verge, 15km south of Katanning

SE 1
Fringed Mantis Orchid (Oct)
Rattle Beaks (Sep)
Stark White Spider Orchid (Oct)

SE 2
Bee Orchid (*Diuris*) (Oct)
Stark White Spider Orchid (Oct)
Vanilla Orchid (Sep)

MAP 29

SE 3: Badjebup, Katanning–Nyabing Road

SE 3
Stark White Spider Orchid (Oct)

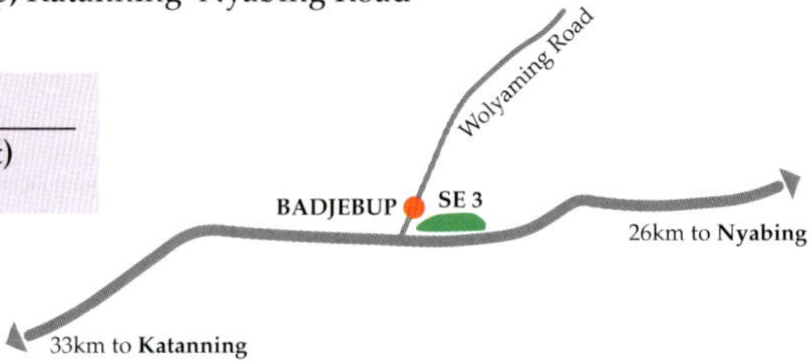

MAP 30

SE 4: Mt Trio area, Stirling Range
SE 5: Verge, south of Stirling Range Retreat on Chester Pass Road

SE 4
Eastern Queen of Sheba (Aug)
Reaching Spider Orchid (Sep)

SE 5
Midge Orchid (Jul)
Stark White Spider Orchid (Oct)

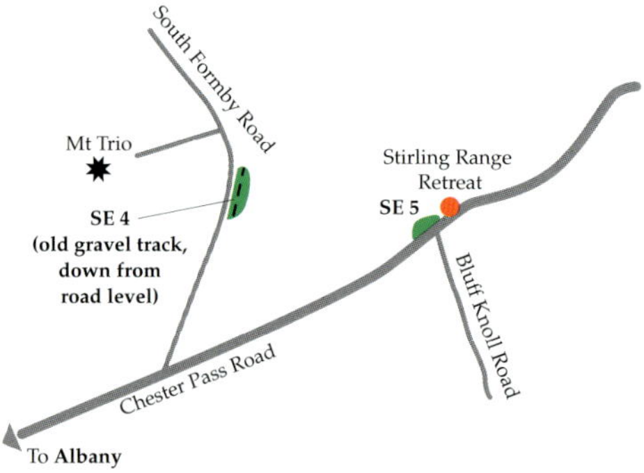

MAP 31

SE 6: Intersection, Toompup South Road–Borden Bremer Road
SE 7: Greaves Hill Road, off Borden Bremer Road
SE 8: Intersection, Nightwell Road South–Borden Bremer Road

SE 6

Dark Banded Greenhood (Aug)
Eastern Queen of Sheba (Aug)
Red Beaks (Aug)

SE 7

Fringed Mantis Orchid (Oct)

SE 8

Common Dragon Orchid (Oct)
Zebra Orchid (Sep)

MAP 32

SE 9: First creek crossing, south of Jerramungup
SE 10: West side of Quiss Road, off South Coast Highway
SE 11: Rest area, 20km east of Jerramungup
on South Coast Highway
SE 12: Fitzgerald National Park

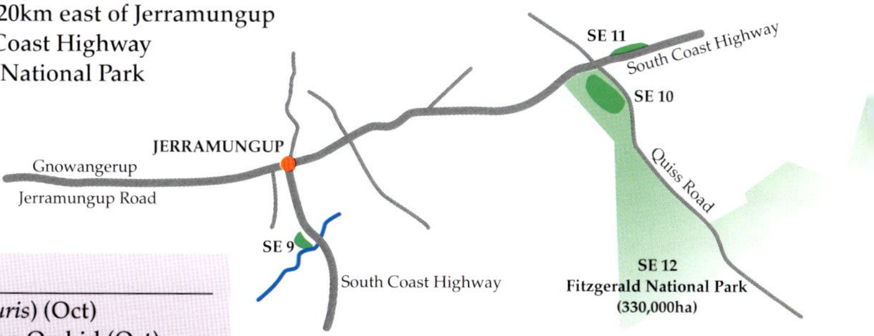

SE 9

Bee Orchid (*Diuris*) (Oct)
Common Dragon Orchid (Oct)
Fringed Mantis Orchid (Oct)
Lazy Spider Orchid (Oct)
Pink Candy Orchid (Aug)
Slender Spider Orchid (Oct)
Stark White Spider Orchid (Oct)
Vanilla Orchid (Sep)
Zebra Orchid (Sep)

SE 10

Blue China Orchid (Sep)
Purple Enamel Orchid (Sep)
Red Beaks (Aug)
White Mignonette Orchid (Oct)

SE 11

Bee Orchid (*Diuris*) (Oct)
Purple Enamel Orchid (Sep)
White Mignonette Orchid (Oct)

SE 12

Blue China Orchid (Sep)
Leopard Orchid (Nov)
Purple Enamel Orchid (Sep)
White Mignonette Orchid (Oct)

MAP 33

SE 13: Carlingup Road, Ravensthorpe, off South Coast Highway

SE 13
Bee Orchid (*Diuris*) (Oct)
Pink Candy Orchid (Aug)
Small Mantis Orchid (Sep)
Western Wispy Spider Orchid (Sep)

MAP 34

SE 14: Rotary Lookout, Esperance
SE 15: Helms Arboretum, Esperance, off Coolgardie Esperance Highway

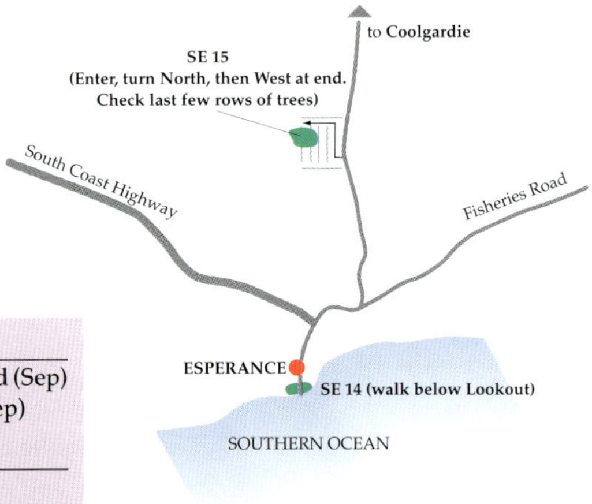

SE 14
Esperance White Spider Orchid (Sep)
Rigid White Spider Orchid (Sep)

SE 15
Bee Orchid (*Diuris*) (Oct)
Cowslip Orchid (Sep)
Esperance King Spider Orchid (Sep)
Heberle's Spider Orchid (Sep)

MAP 35

SE 16: Boyatup Hill, 114km east of Esperance, off Fisheries Road

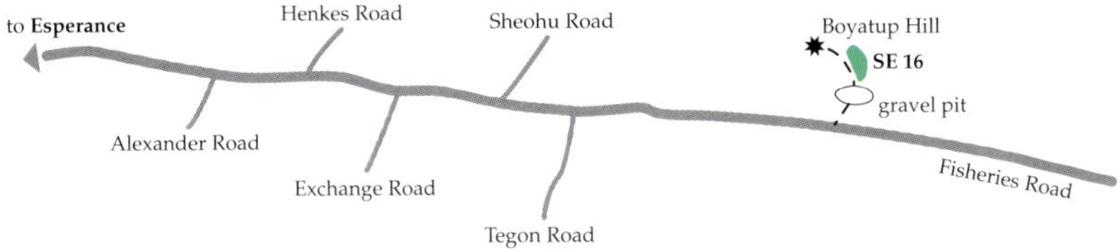

SE 16

Beautiful Donkey Orchid (Sep)
Bee Orchid (*Diuris*) (Oct)
Esperance King Spider Orchid (Sep)
Heberle's Spider Orchid (Sep)
Midge Orchid (Jul)
Pink Candy Orchid (Aug)
Pointing Spider Orchid (Sep)
Purple Enamel Orchid (Sep)
Rabbit Orchid (Sep)
Rattle Beaks (Sep)
Vanilla Orchid (Sep)
White Fairy Orchid (Sep)
Zebra Orchid (Sep)

MAP 36

SE 17: Parking bay/gravel scrape between Lake Grace and Newdegate
SE 18: South Buniche Nature Reserve, near sign, ~1km east of SE17

SE 17

Chameleon Orchid (Sep)
Jug Orchid (Sep)
Pouched Leek Orchid (Oct)

SE 18

Chameleon Orchid (Sep)
Jug Orchid (Sep)

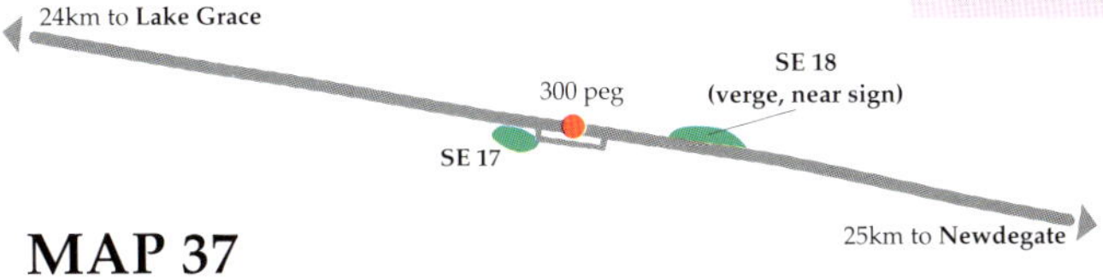

MAP 37

SE 19: Harrismith area

SE 19

Chameleon Orchid (Sep)
Dark Banded Greenhood (Aug)
Jug Orchid (Sep)
Purple Enamel Orchid (Sep)

MAP 38

E 1: Canning Road end of Welshpool Road, Carmel

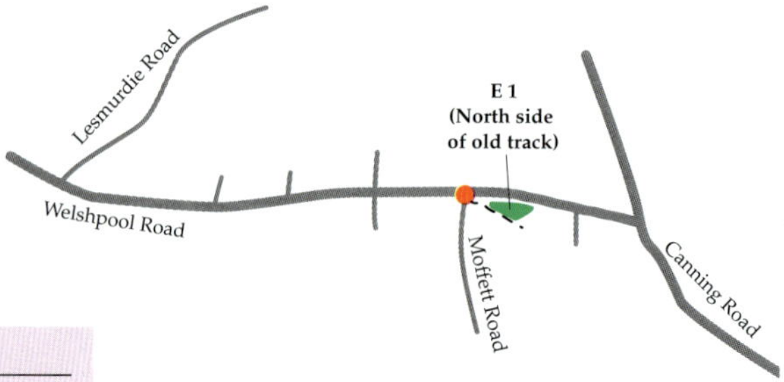

E 1

Hare Orchid (May)

MAP 39

E 2: Yarra Road and Wandoo Conservation Park

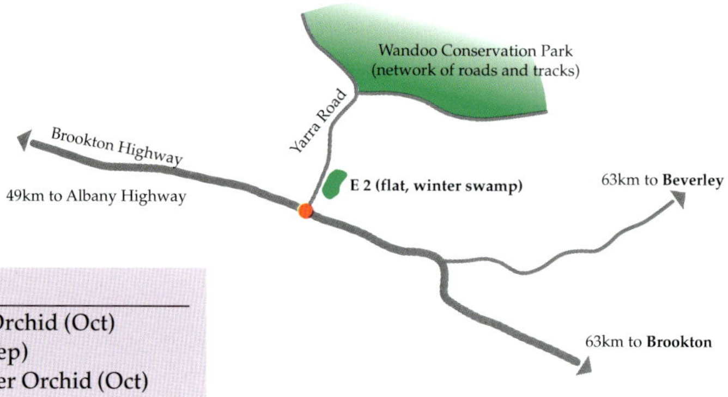

E 2

Fringed Mantis Orchid (Oct)
Rabbit Orchid (Sep)
Stark White Spider Orchid (Oct)

MAP 40

E 3: Sorenson's Reserve, Babakin

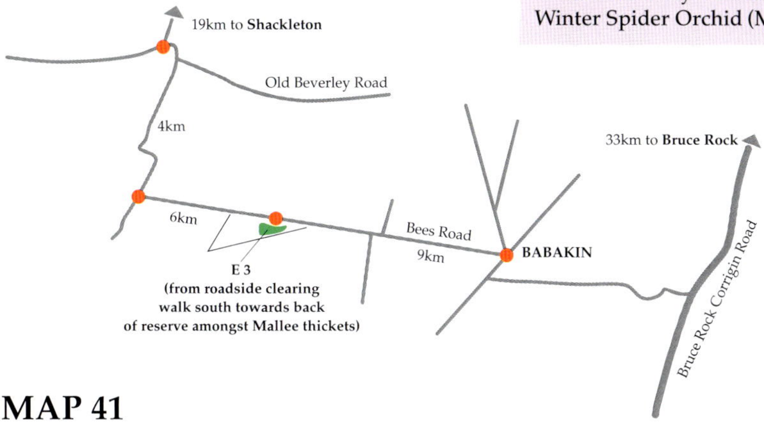

E 3

Crinkle-leafed Bunny Orchid (May)
Common Bunny Orchid (May)
Winter Spider Orchid (May)

MAP 41

E 4: Reserve off Goomalling–Toodyay Road

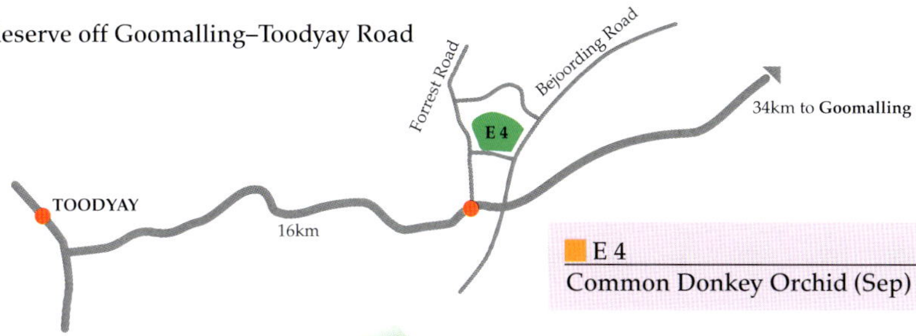

E 4

Common Donkey Orchid (Sep)

MAP 42

E 5: Avon National Park

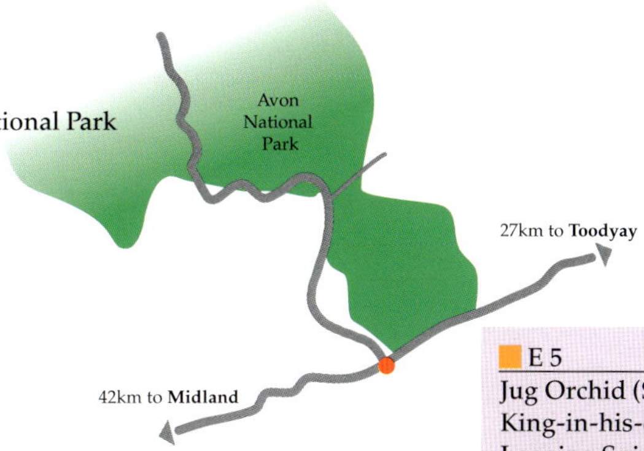

E 5

Jug Orchid (Sep)
King-in-his-carriage (Sep)
Leaping Spider Orchid (Sep)
Pink Enamel Orchid (Oct)
Zebra Orchid (Sep)

MAP 43

N 1: Neerabup National Park

N 1
- Coastal White Spider Orchid (Sep)
- Purple Enamel Orchid (Sep)
- Shirt Orchid (Sep)

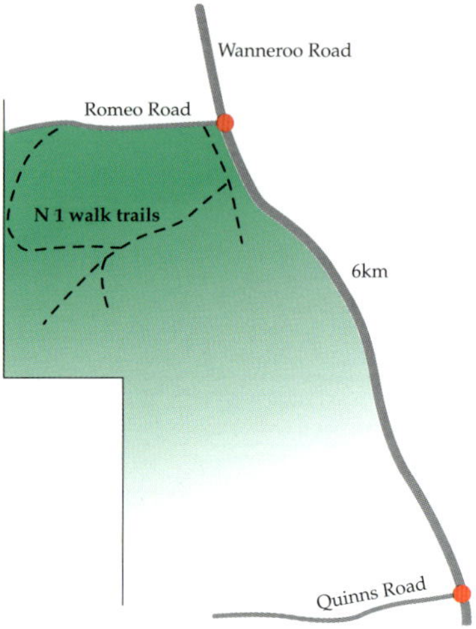

MAP 44

N 2: Regans Ford Area

N 2
- Arrowsmith Pansy Orchid (Aug)
- Blue Fairy Orchid (Aug)
- Cowslip Orchid (Sep)
- Noble Spider Orchid (Sep)

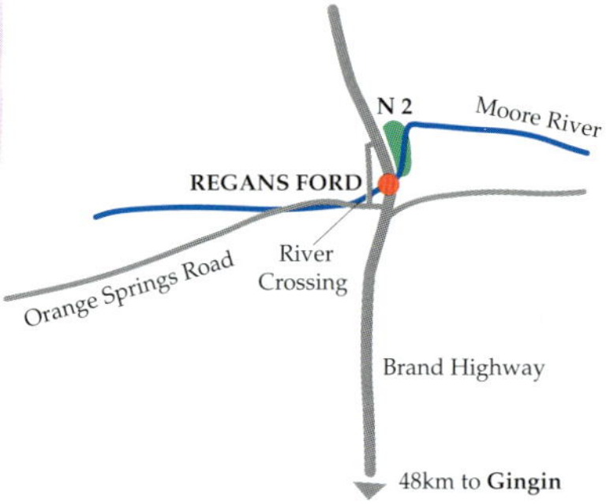

MAP 45

N 3: Gravel pit south of Cataby, off Brand Highway

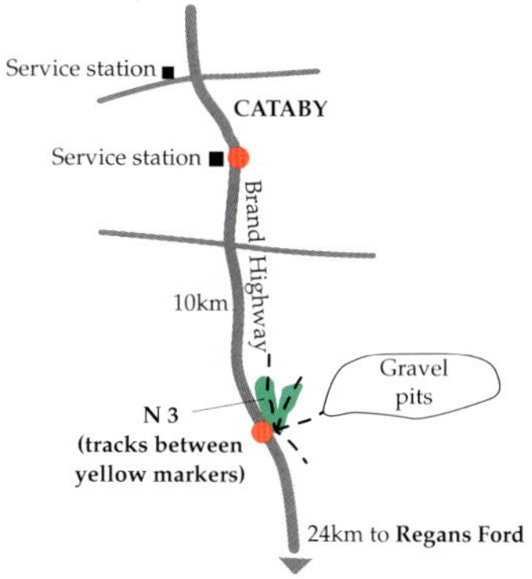

N 3
Coastal Banded Greenhood (Jul)
Dark Banded Greenhood (Aug)
Hare Orchid (May)

MAP 46

N 4: Badgingarra

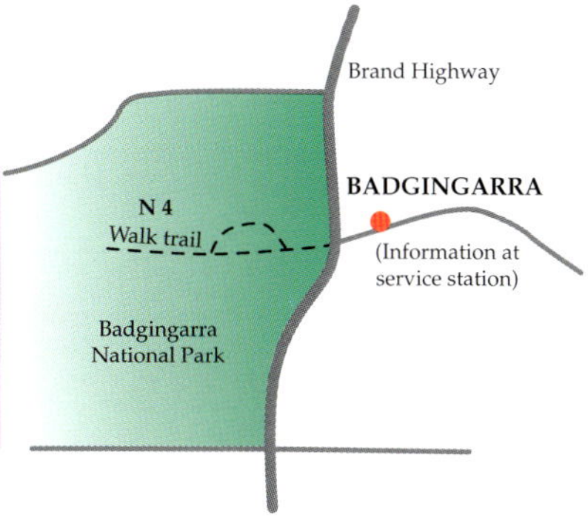

N 4
Blue Fairy Orchid (Aug)
Common Donkey Orchid (Sep)
Common Spider Orchid (Aug)
Little Pink Fairy Orchid (Sep)
Purple Enamel Orchid (Sep)

MAP 47

N 5: Wandoo Reserve, Jurien Road

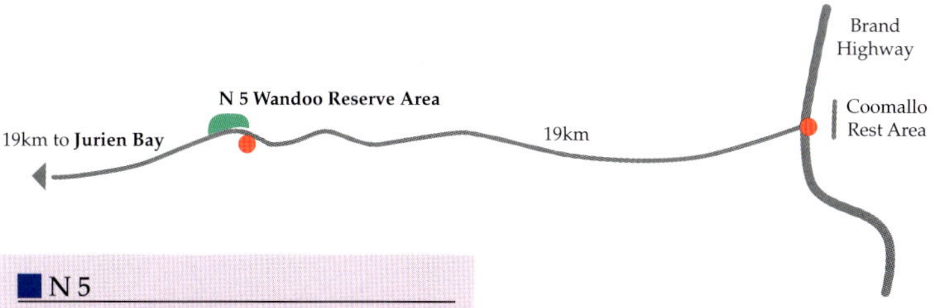

N 5
Daddy-long-legs Spider Orchid (Aug)
Blue Fairy Orchid (Aug)

MAP 48

N 6: Verge, 1km south of Eneabba, Brand Highway
N 7: Erindoon Road, off Coolinda–Eneabba Road
N 8: Verge, Coolinda–Eneabba Road
N 9: Old gravel scrape, off Coolinda–Eneabba Road
N 10: Lake Indoon

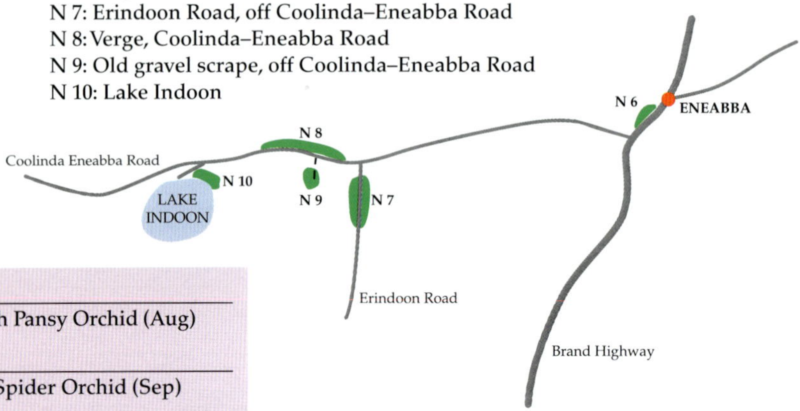

N 6
Arrowsmith Pansy Orchid (Aug)

N 7
Drooping Spider Orchid (Sep)

N 8
Arrowsmith Pansy Orchid (Aug)
Blue Fairy Orchid (Aug)
Cowslip Orchid (Sep)
Drooping Spider Orchid (Sep)

N 9
Daddy-long-legs Spider Orchid (Aug)
Queen of Sheba (Aug)

N 10
Daddy-long-legs Spider Orchid (Aug)

MAP 49

N 11: Western Flora, Brand Highway
N 12: Arrowsmith turn off from Brand Highway

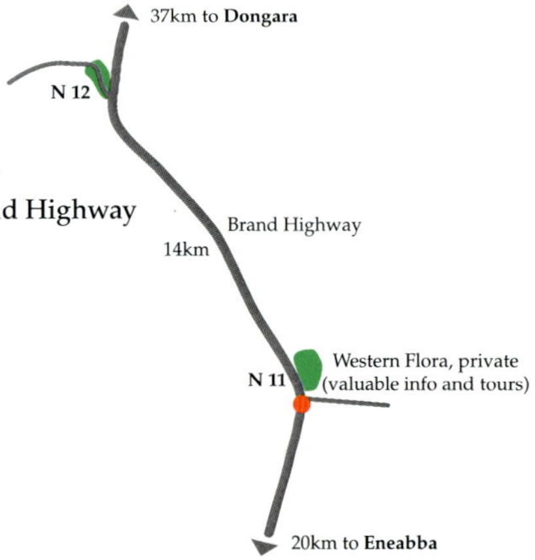

N 11
more than 60 species

N 12
Arrowsmith Spider Orchid (Aug)
Noble Spider Orchid (Sep)

MAP 50

N 13: End of Swamp Road, Northampton

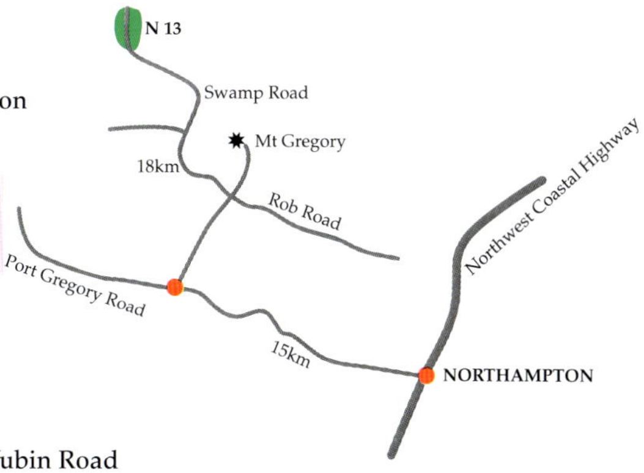

N 13
Kalbarri Cowslip Orchid (Sep)

MAP 51

N 14: Verge, Mullewa–Wubin Road
N 15: Verge, Tardun–Pindar Road
N 16: Wildflower Reserve, Mullewa–Wubin Road
N 17: Canna, Nature Reserve

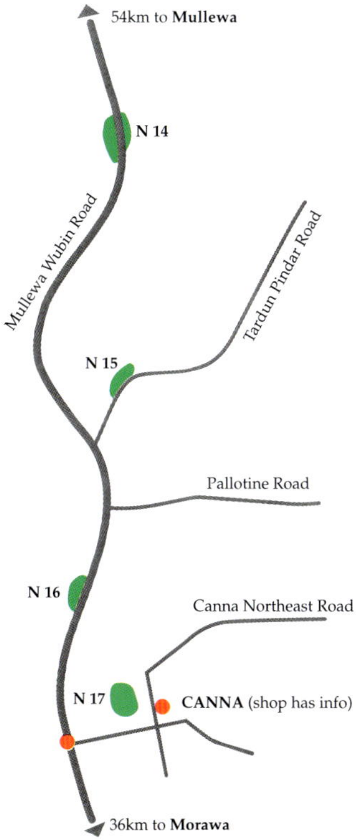

N 14
Ant Orchid (Sep)
Blue Fairy Orchid (Aug)
Rosy-cheeked Donkey Orchid (Sep)

N 15
Ant Orchid (Sep)
Hairy-stemmed Snail Orchid (Aug)

N 16
Blue Fairy Orchid (Aug)
Dainty Donkey Orchid (Sep)
Hairy-stemmed Snail Orchid (Aug)
Outback Spider Orchid (Aug)

N 17
Ant Orchid (Sep)
Blue Fairy Orchid (Aug)
Dainty Donkey Orchid (Sep)
Hairy-stemmed Snail Orchid (Aug)
Pansy Orchid (Sep)
Pink Candy Orchid (Aug)
Rosy-cheeked Donkey Orchid (Sep)
Rothsay Spider Orchid (Aug)
Yawning Leek Orchid (Sep)

MAP 52

N 18: Coalseam Conservation Park
N 19: Depot Hill

N 18
Ant Orchid (Sep)
Chameleon Orchid (Sep)
Pink Candy Orchid (Aug)
Spoon-lipped Rufous Greenhood (Sep)

N 19
Blue Fairy Orchid (Aug)
Pink Candy Orchid (Aug)

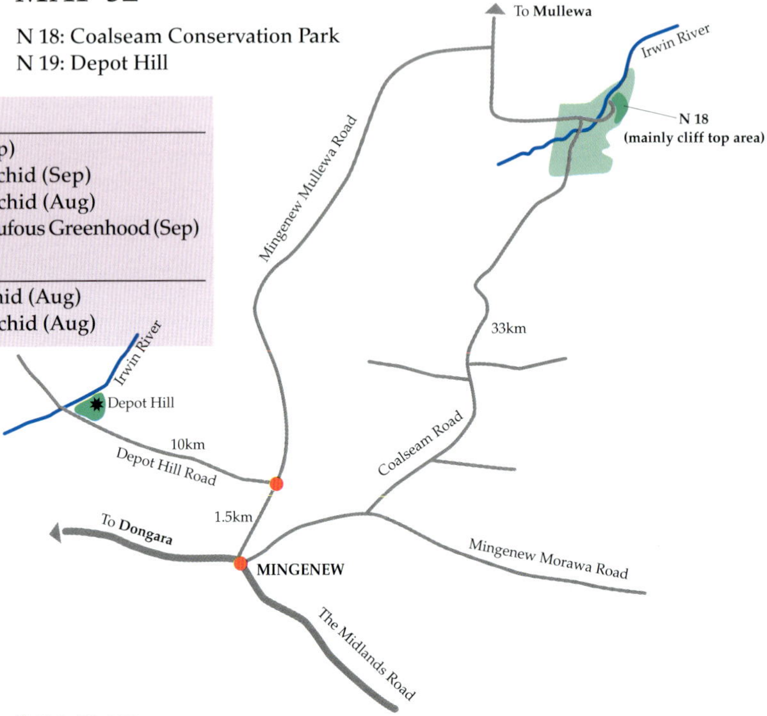

MAP 53

N 20: Morawa Township

N 20
Chameleon Orchid (Sep)
Hairy-stemmed Snail Orchid (Aug)

MAP 54

N 21: Wubin

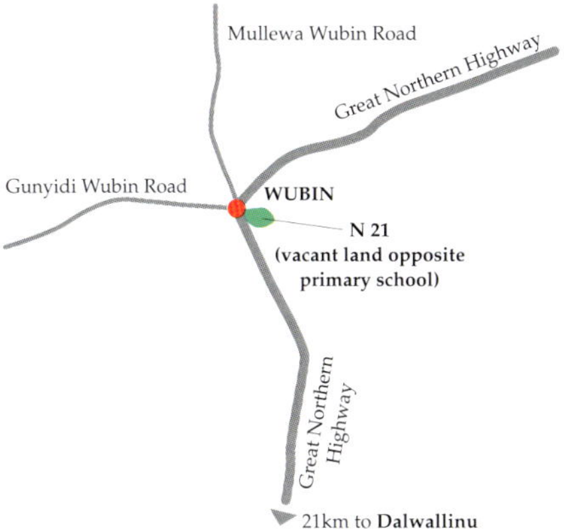

N 21
Chameleon Orchid (Sep)
Dainty Donkey Orchid (Sep)

MAP 55

N 22: Reserve, Kiaka Road, north of Moora

N 22
Blue Fairy Orchid (Aug)
Pink Candy Orchid (Aug)

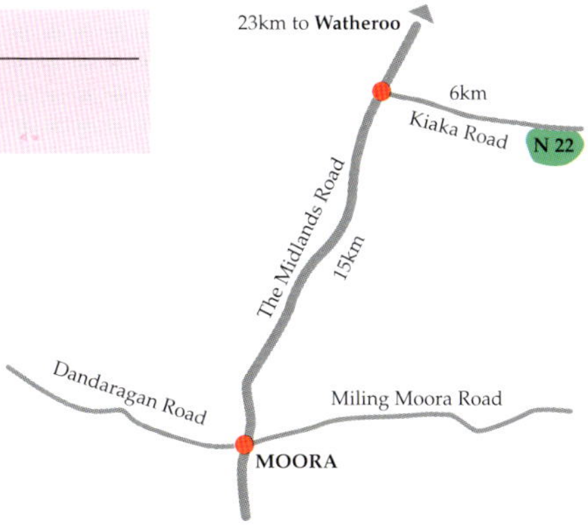

Indexes

Index to Scientific Names

Index to Common Names

Associated Plant Species

COMMON NAMES		SCIENTIFIC NAMES
Balga (Blackboy)		*Xanthorrhoea preissii*
Banksia	various	*Banksia* species
Candle Banksia		*B. attenuata*
Bull Banksia		*B. grandis*
Swamp Banksia		*B. littoralis*
Illyarrie		*Eucalyptus erythrocorys*
Jam		*Acacia acuminata*
Jarrah		*Eucalyptus marginata*
Karri		*Eucalyptus diversicolor*
Mallee	various	*Eucalyptus* species
Marri (Redgum)		*Corymbia calophylla*
Paperbark	various	*Melaleuca* species e.g. *rhaphiophylla* (Swamp paperbark)
Peppermint		*Agonis flexuosa*
Sheoak	various	*Allocasuarina* species e.g. *fraseriana*
Tuart		*Eucalyptus gomphocephala*
Wandoo		*Eucalyptus wandoo*
Yate		*Eucalyptus occidentalis*
York gum		*Eucalyptus loxophleba*
Zamia palm		*Macrozamia riedlei*

Maps are not to scale and for general guide purposes only
PERTH
Rockingham
Mandurah
Pinjarra
INDIAN OCEAN
SW
(South West)
Bunbury
Collie
Capel
Donnybrook
Cape Naturaliste
Dunsborough
Busselton
Carbunup
Yallingup
Gracetown
Margaret River
Witchcliffe
Karridale
Augusta
Cape Leeuwin
Nannup
Boyup Brook
Kojonup
Manjimup
Rocky Gully
1
2
3
4
5
6
7
8
9
10
11
12
13
14-16
17
18
19-21
22
23-37
38
39
40
41
42
43
44